Charged Up: A Comprehensive Guide to Electric Vehicles

by

Etienne Psaila

Charged Up: A Comprehensive Guide to Electric Vehicles

Copyright © 2024 by Etienne Psaila. All rights reserved.

First Edition: **April 2024**

No part of this publication may be reproduced, distributed, or transmitted in any form or by any means, including photocopying, recording, or other electronic or mechanical methods, without the prior written permission of the publisher, except in the case of brief quotations embodied in critical reviews and certain other non-commercial uses permitted by copyright law.

This book is part of the 'Automotive and Motorcycle Books' series and each volume in the series is crafted with respect for the automotive and motorcycle brands discussed, utilizing brand names and related materials under the principles of fair use for educational purposes. The aim is to celebrate and inform, providing readers with a deeper appreciation for the engineering marvels and historical significance of these iconic brands.

Cover design by Etienne Psaila
Interior layout by Etienne Psaila

Website: **www.etiennepsaila.com**
Contact: **etipsaila@gmail.com**

Table of Contents

Chapter 1: Introduction to Electric Vehicles

Chapter 2: Early Inventions and Developments

Chapter 3: The Golden Age of Electric Vehicles

Chapter 4: Decline and Resurgence

Chapter 5: Technological Advances and Market Growth

Chapter 6: Challenges and Criticisms

Chapter 7: EVs Today and the Road Ahead

Chapter 8: Manufacturing

Chapter 9: Batteries

Chapter 10: Real Impact on the Environment

Chapter 11: Charging and Infrastructure

Chapter 12: Legislation

Chapter 13: Your Guide to Buying an Electric Vehicle in 2024

Chapter 14: Future Trends and Technology

Chapter 1: Introduction to Electric Vehicles

Definition and Overview

Electric vehicles (EVs), at their core, are vehicles propelled by electric motors, using energy typically stored in rechargeable batteries. Unlike conventional vehicles that use gasoline or diesel-powered engines, EVs operate on electricity, varying from cars and buses to bikes and scooters. The concept of electric mobility is not new; it dates back to the 19th century, when the first prototypes and models hinted at the possibility of transportation without the need for combustible fuels.

The appeal of EVs lies not just in their operational efficiency or low maintenance requirements but in their potential to revolutionize our transportation systems towards sustainability. As the world grapples with climate change and air quality concerns, EVs offer a cleaner, more environmentally friendly alternative to traditional internal combustion engine vehicles, by significantly reducing greenhouse gas emissions and pollution.

The Importance of EVs

The significance of electric vehicles transcends environmental benefits. EVs are at the forefront of technological innovation in the automotive industry, integrating advancements in battery technology, autonomous driving, and connectivity. Their rise is synonymous with a broader shift towards sustainable development, underpinned by global efforts to reduce carbon emissions and combat climate change.

Moreover, the push for electric mobility is increasingly supported by governments around the world through incentives, infrastructure development, and stringent emissions regulations, reflecting a collective commitment to a cleaner, greener future. For consumers, EVs offer the promise of lower operating costs, new driving experiences, and the opportunity to partake in the burgeoning ecosystem of sustainable technologies.

As we embark on this journey through the history and development of electric vehicles, it's essential to recognize their role not just as a mode of transportation but as a catalyst for change. By understanding where EVs have come from and envisioning where they are headed, we can appreciate the profound impact they are poised to have on our world.

Chapter 2: Early Inventions and Developments

The journey of electric vehicles (EVs) from mere curiosity to mainstream acceptance is a fascinating story of innovation, rivalry, and vision. This section explores the dawn of electric mobility, highlighting the key inventions and developments that paved the way for the EVs we know today.

Origins of the Electric Vehicle

The concept of electric transportation can be traced back to the early 19th century. The first experimental electric cars were built in the 1830s and 1840s. In 1834, Thomas Davenport, an American inventor, constructed a small-scale electric car powered by a non-rechargeable battery. Across the Atlantic, Scottish inventor Robert Anderson developed a similar prototype around the same period. These early models were rudimentary, demonstrating the potential of electric propulsion but lacking in practicality and efficiency.

Innovation and Experimentation

The late 19th century witnessed significant advancements in battery technology, which, in turn, propelled the development of more practical and powerful electric vehicles. In 1859, French physicist Gaston Planté invented the lead-acid battery, the first rechargeable battery, which became a foundational technology for electric vehicles.

The 1880s and 1890s were marked by a flurry of innovation. Notable figures such as Thomas Edison and Nikola Tesla explored electric vehicles and their components, seeking to improve efficiency and performance. Edison, in particular, worked on developing a more durable battery, believing that electric cars were the superior technology.

Commercial Success and Public Interest

The first commercially available electric car was introduced in 1888 by the German engineer Andreas Flocken. Dubbed the Flocken Elektrowagen, this vehicle is often considered the first real electric car. In the following decade, electric vehicles began to gain popularity, especially in urban areas in the United States and Europe. They were marketed as clean, quiet, and easy to drive, in stark contrast to their noisy and smelly gasoline counterparts.

In 1897, electric vehicles made their mark on public transportation with the launch of the Electric Carriage and Wagon Company in London, and the New York Electric Vehicle Transportation Company in the United States, introducing electric taxis to the streets of New York City. The early 20th century saw electric vehicles set numerous speed and distance records, showcasing their potential to compete with, and even outperform, gasoline vehicles.

Key Models and Innovations

Several electric vehicle models from this era left a significant mark on the history of EVs. The Detroit Electric, produced by the Anderson Electric Car Company from 1907 to 1939, became one of the most popular electric cars, renowned for its reliability and range, which could exceed 80 miles on a single charge. Another notable model was the Baker Electric, which boasted famous owners like Thomas Edison.

This period of early inventions and developments laid the groundwork for the future of electric mobility. It was characterized by a wave of enthusiasm and optimism for electric vehicles, setting the stage for their evolution. However, as the 20th century progressed, the advent of cheaper and more accessible gasoline-powered vehicles gradually overshadowed electric cars, leading to their decline. Despite this setback, the foundational work of early inventors and the initial popularity of EVs played a crucial role in keeping the vision of electric mobility alive.

Chapter 3: The Golden Age of Electric Vehicles

At the turn of the 20th century, electric vehicles (EVs) experienced what many consider their "Golden Age." This period, spanning the late 1890s to the early 1920s, saw EVs emerge as not only viable but preferable modes of transportation for a growing urban population. Amidst burgeoning cities and evolving societal needs, electric vehicles found their niche, appealing to consumers with their ease of use, reliability, and cleanliness.

Popularity and Advantages

The allure of electric vehicles during this era stemmed from several key advantages over their gasoline-powered counterparts. Firstly, EVs were simpler to operate, lacking the manual effort required to start gasoline engines. They were also quieter and did not emit the noxious fumes that were characteristic of gasoline vehicles, making them particularly appealing to urban dwellers who valued cleanliness and noise-free operation.

Electric vehicles became the vehicle of choice for many city commuters, as well as for women drivers, who appreciated the lack of a crank start and the smoother ride. These attributes contributed to the image of electric vehicles as the sophisticated choice for the modern urbanite.

Notable Models and Milestones

During the Golden Age, several electric vehicle models captured the public's imagination and demonstrated the potential of electric mobility. The Detroit Electric, produced by the Anderson Electric Car Company, became synonymous with reliability and comfort, boasting a single-charge range that rivaled, and sometimes surpassed, that of gasoline vehicles.

Another prominent figure of this era was the Baker Electric. Marketed with a focus on female drivers, it featured luxurious interiors and custom body styles, setting a precedent for the customization and personalization of vehicles. The Baker Electric's clientele included influential figures such as Thomas

Edison and John D. Rockefeller, Jr., underscoring the status associated with electric vehicles.

The golden age also saw the emergence of electric trucks and buses, which were used in cities for deliveries and public transportation. These vehicles demonstrated the versatility of electric powertrains and their applicability to a range of transportation needs.

Challenges and the Beginning of Decline

Despite their early success, electric vehicles began facing challenges as the 1920s progressed. The discovery of large oil reserves led to a dramatic decrease in the price of gasoline, making internal combustion engine vehicles cheaper to operate. At the same time, technological advancements such as the electric starter motor made gasoline cars much easier to use, eroding one of the key advantages of electric vehicles.

Furthermore, the expanding network of roads connecting cities allowed people to travel longer distances, highlighting the limitations in the range of electric vehicles compared to gasoline-powered cars. This shift in consumer preference towards vehicles that could travel further without the need for lengthy recharges contributed to the decline in the popularity of EVs.

Legacy of the Golden Age

Though the golden age of electric vehicles came to an end, the era left an indelible mark on the automotive industry. It demonstrated the practicality and desirability of electric mobility and set the stage for the eventual resurgence of interest in EVs. The innovations in vehicle design and technology during this period laid the groundwork for future developments in electric transportation.

The legacy of the golden age is a testament to the enduring appeal of electric vehicles, their impact on early automotive culture, and their potential to meet the needs of urban populations. As we look back on this pivotal period, it becomes clear that the journey of electric vehicles was not linear but a cycle of discovery, innovation, and renaissance.

Chapter 4: Decline and Resurgence

The mid-20th century marked a period of decline for electric vehicles (EVs), as they gradually faded from the forefront of automotive innovation and public interest. This decline was not a sudden fall but a gradual eclipse by gasoline-powered vehicles, which offered greater range, convenience, and affordability. However, the story of EVs did not end there. The latter part of the 20th century and the early 21st century witnessed a remarkable resurgence, fueled by technological advancements, environmental awareness, and shifting consumer preferences.

Reasons for Decline

Several factors contributed to the decline of electric vehicles:

- **Advancements in Internal Combustion Engine (ICE) Vehicles**: Innovations such as the electric starter motor, introduced in 1912, eliminated the need for a hand crank, making gasoline cars much easier to start. Improved engineering and mass production techniques, notably by Henry

Ford with the Model T, significantly lowered the cost of ICE vehicles.

- **Expansion of Oil Infrastructure**: The discovery of vast oil reserves, particularly in Texas and the Middle East, led to a dramatic decrease in the price of gasoline, making ICE vehicles cheaper to operate.

- **Increased Range and Infrastructure**: The development of a sprawling road infrastructure enabled longer-distance travel, highlighting one of the electric vehicle's main limitations: its range. Gasoline vehicles, with their extended range and rapidly growing network of filling stations, offered a level of freedom and convenience that electric vehicles could not match at the time.

The Resurgence in Interest

The resurgence of electric vehicles can be attributed to a confluence of factors:

- **Environmental Concerns**: The oil crises of the

1970s, along with growing awareness of environmental issues such as air pollution and climate change, prompted both consumers and governments to reconsider the sustainability of transportation.

- **Technological Advancements**: Significant improvements in battery technology, notably the development of lithium-ion batteries, have dramatically increased the range and decreased the charging time of EVs. Advances in power electronics and motor technology have also contributed to the efficiency and performance of electric vehicles.

- **Government Policies and Incentives**: Many governments around the world introduced policies to promote the adoption of electric vehicles, motivated by environmental concerns and the desire to reduce dependence on imported oil. These policies include tax incentives, subsidies, and mandates for the production and sale of EVs, as well as investments in charging infrastructure.

Key Milestones in the Resurgence

- **The Introduction of the Toyota Prius**: Although not a fully electric vehicle, the launch of the Toyota Prius in 1997 as the world's first mass-produced hybrid electric vehicle marked a significant step toward the acceptance of electrified transportation.

- **The Rise of Tesla Motors**: Founded in 2003, Tesla Motors (now Tesla, Inc.) played a pivotal role in changing the perception of electric vehicles from niche to mainstream. The Tesla Roadster, introduced in 2008, was the first highway-legal serial production all-electric car to use lithium-ion battery cells. Tesla's subsequent models, including the Model S, have continued to push the boundaries of EV technology and appeal.

- **Mainstream Automakers Embrace EVs**: Following Tesla's success, established automakers began to seriously invest in electric vehicle development, leading to a broad range of EVs across various market segments. Notable launches include the Nissan Leaf in 2010, the Chevrolet Bolt in 2016, and the proliferation of electric models from European luxury brands.

The Road Ahead

The resurgence of electric vehicles represents a convergence of vision, innovation, and necessity. As we look to the future, the trajectory of EVs is increasingly clear, with electric mobility central to the global strategy for sustainable development, climate change mitigation, and technological progress. The decline and subsequent resurgence of electric vehicles are not merely a chapter in automotive history but a reflection of broader societal shifts towards sustainability and innovation.

Chapter 5: Technological Advances and Market Growth

The resurgence of electric vehicles (EVs) is not just a tale of increasing environmental awareness and policy support; it's also a story of remarkable technological innovation and market expansion. As we entered the 21st century, several key technological advances underpinned the growth of the EV market, transforming electric vehicles from niche products into mainstream solutions.

Breakthroughs in Battery Technology

The heart of an electric vehicle's innovation lies in its battery technology. Lithium-ion batteries, in particular, have been pivotal in this transformation, offering higher energy density, longer lifespans, and faster charging times compared to previous battery technologies. This advancement has directly impacted the range and usability of EVs, making them a practical option for a broader range of consumers.

- **Energy Density**: The increase in energy density

has allowed batteries to store more power without significantly increasing in size or weight, extending the range of EVs to rival that of gasoline vehicles.

- **Charging Time**: Developments in charging technology, including fast-charging stations, have dramatically reduced the time required to recharge an EV, addressing one of the key concerns of potential EV owners.

- **Cost Reduction**: Economies of scale, driven by increased production and technological improvements, have lowered the cost of lithium-ion batteries. This cost reduction is crucial in making EVs competitive with internal combustion engine vehicles in terms of price.

Advancements in Electric Motors and Power Electronics

Electric motors have also seen significant advancements, with improvements in efficiency, size, and power. The integration of power electronics, which control the flow of electricity between the battery and the motor, has enhanced the performance and efficiency of EVs. These technological advancements have enabled electric vehicles to offer competitive acceleration and top speeds.

Integration of Renewable Energy

The growth of the EV market is closely linked with the integration of renewable energy sources. As the electricity grid becomes greener, the environmental benefits of electric vehicles increase. The ability to charge EVs with solar or wind energy further reduces the carbon footprint of electric mobility, aligning with global sustainability goals.

Government Policies and Market Incentives

Government initiatives have played a crucial role in accelerating the adoption of electric vehicles. Incentives such as tax rebates, grants, and subsidies for both manufacturers and consumers have made EVs more financially accessible. Moreover, investments in charging infrastructure and mandates for EV sales have supported the market's growth.

- **Global EV Sales Trends**: The impact of these technological and policy advancements is evident in the global sales trends of electric vehicles. EV sales have been growing exponentially, with countries like Norway, China, and the United States leading the charge in adoption rates.

- **Automotive Industry's Shift**: Recognizing the potential and responding to regulatory pressures, major automakers have committed to electrifying their fleets. The industry has seen a significant shift, with billions being invested in EV development and production.

The Road Ahead

The technological advancements and market growth of electric vehicles signal a paradigm shift in transportation. As battery technology continues to evolve and charging infrastructure expands, the remaining barriers to EV adoption are rapidly diminishing. The future of the automotive industry is electric, with EVs playing a central role in the transition towards sustainable mobility.

The increasing variety of EV models, from sedans and SUVs to trucks and buses, reflects the expanding appeal of electric mobility. With ongoing innovation and supportive policies, the market for electric vehicles is set to continue its upward trajectory, reshaping how we think about transportation, energy, and the environment.

Chapter 6: Challenges and Criticisms

While the ascent of electric vehicles (EVs) represents a significant shift towards sustainable transportation, the journey is not without its hurdles. Several challenges and criticisms have emerged, reflecting the complexities of transitioning to a predominantly electric vehicle fleet. Addressing these concerns is crucial for the continued growth and acceptance of EVs.

Range Anxiety

One of the most persistent concerns associated with electric vehicles is range anxiety—the fear that an EV does not have enough battery range to reach its destination, leaving the driver stranded without charging options. Despite significant improvements in battery technology that have increased the range of many new EV models to over 300 miles per charge, the specter of range anxiety still affects potential buyers, particularly those without access to home charging or those who regularly undertake long-distance travel.

Charging Infrastructure

Closely related to range anxiety is the challenge of developing a comprehensive and accessible charging infrastructure. While urban areas and certain regions have seen a significant increase in charging stations, many areas, especially rural and underserved regions, lack sufficient public charging options. The disparity in charging access can be a significant barrier to EV adoption, requiring coordinated efforts from governments, businesses, and communities to address.

Battery Production and Recycling

The environmental impact of producing and recycling EV batteries has also come under scrutiny. The mining of lithium, cobalt, and other raw materials necessary for battery production raises concerns about ecological damage, water use, and human rights issues in mining communities. Furthermore, the recycling of EV batteries remains a challenge, with a need for more efficient and effective recycling processes to recover valuable materials and reduce landfill waste.

Energy Source and Carbon Footprint

Critics also point out that the environmental benefits of electric vehicles are contingent on the source of electricity used to charge them. In regions where the electricity grid relies heavily on fossil fuels, the net reduction in greenhouse gas emissions from EVs may be less significant. This criticism underscores the importance of transitioning to renewable energy sources alongside the adoption of electric vehicles to maximize their environmental benefits.

Economic and Social Equity

The transition to electric vehicles has raised questions about economic and social equity. The initial cost of EVs remains higher than that of comparable gasoline vehicles, although total ownership costs over time can be lower due to savings on fuel and maintenance. Additionally, there are concerns about the impact of the transition on workers in industries related to traditional automotive manufacturing and fossil fuel production. Ensuring that the benefits of electric mobility are accessible to all segments of society and that workers affected by the transition have opportunities for retraining and employment in emerging industries is a critical consideration.

Responding to Challenges and Criticisms

Addressing these challenges requires a multifaceted approach that includes continued technological innovation, strategic policy interventions, and collaborative efforts across sectors. Investment in research and development can lead to further improvements in battery technology, reducing costs and environmental impacts. Expanding and enhancing charging infrastructure, particularly in underserved areas, can alleviate range anxiety and make EVs a more viable option for a broader range of consumers. Meanwhile, policies and programs that promote renewable energy, equitable access to electric mobility, and support for transitioning workers can address broader environmental and social equity concerns.

Chapter 7: EVs Today and the Road Ahead

The landscape of electric vehicles (EVs) today is markedly different from any previous point in history. With soaring adoption rates, rapid technological advancements, and shifting consumer attitudes, EVs are not just a part of the automotive future; they are shaping the present. This section examines the current state of the EV market and speculates on the future developments that will drive the evolution of electric mobility.

The Current State of EVs

- **Adoption Rates**: The global adoption of electric vehicles has been accelerating at an unprecedented pace. Countries like Norway, China, and the Netherlands are leading the way, with EVs making up a significant portion of new car sales. This surge is driven by a combination of factors, including government incentives, increasing consumer awareness of environmental issues, and the growing availability of EV models across different price points and segments.

- **Technology Trends**: Technological innovations continue to drive the evolution of the EV market. Advancements in battery technology are increasing energy density, reducing charging times, and lowering costs. Electric drivetrains are becoming more efficient, and integration with digital and autonomous driving technologies is enhancing the user experience. Moreover, the development of solid-state batteries and improvements in charging infrastructure, including wireless charging and ultra-fast charging stations, are on the horizon.

- **Consumer Attitudes**: The perception of electric vehicles has shifted significantly. Once viewed as a niche product with limited range and appeal, EVs are now considered mainstream, desirable options. Consumer concerns about range anxiety and charging infrastructure are being addressed, and the environmental benefits of EVs are widely recognized. The growing interest in sustainability and green technology is further bolstering the appeal of electric vehicles.

Future Prospects

- **Battery Technology Advancements**: The next wave of advancements in battery technology, including the commercialization of solid-state batteries, promises to further increase the range of EVs, reduce charging times, and enhance safety. Research into alternative battery chemistries could reduce reliance on scarce materials, addressing both environmental and supply chain concerns.

- **Integration with Renewable Energy**: The future of electric vehicles is closely tied to the transition towards renewable energy. The increasing use of solar and wind power to charge EVs will enhance their environmental benefits. Innovations such as vehicle-to-grid (V2G) technology, which allows EVs to return energy to the grid, will integrate electric vehicles into the broader energy ecosystem.

- **Autonomous and Connected Vehicles**: The convergence of electrification and autonomous driving technology will redefine the automotive experience. Autonomous electric vehicles could significantly impact urban mobility, reducing congestion and emissions, and transforming the concept of vehicle ownership. The integration of EVs with smart city infrastructure and the Internet of Things (IoT) will lead to more efficient and sustainable transportation systems.

- **Government Policy and Market Incentives**: The role of government policy in shaping the future of electric mobility cannot be overstated. Continued support in the form of incentives, infrastructure investment, and regulatory frameworks will be critical in driving the adoption of EVs. Policies aimed at reducing greenhouse gas emissions, such as phasing out internal combustion engine vehicles, will further accelerate the shift to electric mobility.

The Road Ahead

As we look to the future, the trajectory of electric vehicles is both exciting and uncertain. Technological innovations, policy decisions, and shifts in consumer behavior will all play a role in determining the pace and direction of the transition to electric mobility. What is clear, however, is that EVs will continue to play a pivotal role in the movement towards sustainable transportation, reshaping our relationship with cars and driving the global agenda on climate change and energy independence.

The road ahead for electric vehicles is paved with challenges, but also with immense opportunities. As we continue to push the boundaries of what is possible, the future of transportation promises to be cleaner, smarter, and more connected than ever before.

Chapter 8: Manufacturing

The manufacturing process of electric vehicles (EVs) represents a significant departure from traditional automotive manufacturing, reflecting not only the unique components and technologies involved in EVs but also the broader environmental and ethical considerations integral to their production. This section explores the nuances of EV manufacturing, comparing it with the production processes of internal combustion engine (ICE) vehicles, and discusses the innovations, challenges, and supply chain dynamics characteristic of the EV industry.

Comparing EV and ICE Vehicle Manufacturing

The distinction in manufacturing between electric vehicles (EVs) and internal combustion engine (ICE) vehicles fundamentally centers on the complexity and construction of their powertrains. The EV powertrain, with its streamlined composition of an electric motor, a battery pack, and power electronics, embodies a leap towards simplification in automotive engineering. This reduction in mechanical complexity is evident when contrasted with the ICE powertrain, which is a more intricate assembly of components including the engine, transmission, fuel injection systems, and exhaust systems. The ICE powertrain's complexity arises from its hundreds of moving parts, each requiring precise manufacturing, assembly, and integration to function seamlessly as a unit. This complexity not only impacts the manufacturing process but also the maintenance and operational aspects of the vehicle over its lifecycle.

In the manufacturing landscape, the simplified design of the EV powertrain translates to streamlined assembly lines. Fewer moving parts mean fewer steps in the assembly process, reducing the time and labor required to build an EV compared to an ICE vehicle. This efficiency can lead to cost savings and a faster production cycle, allowing manufacturers to respond more swiftly to market demands. The simplicity of the EV design also opens up new possibilities for modular and flexible manufacturing approaches, where different models can be produced on the same line with minimal adjustments.

However, the simplicity of the EV powertrain presents its own set of challenges, most notably in the manufacturing of the battery pack. The battery pack is the heart of an electric vehicle, dictating its range, performance, and overall efficiency. Manufacturing a battery pack is a complex, high-precision process that involves assembling hundreds or thousands of individual lithium-ion cells along with the necessary cooling, management, and safety systems. Each of these components must be engineered and assembled to exacting standards to ensure the pack operates efficiently, safely, and within its thermal thresholds.

The complexity of battery production is further amplified by the need for extensive safety testing. Electric vehicle batteries are subject to thermal runaway, a condition where an increase in temperature can lead to a self-sustaining chain reaction resulting in fire or explosion. Preventing thermal runaway necessitates rigorous quality control measures throughout the manufacturing process, from the selection of materials to the final assembly of the battery pack. This includes precise control over the chemical composition of the cells, the integrity of the battery management system (BMS), and the effectiveness of the cooling system. Each of these elements requires careful design, testing, and integration to ensure the safety and reliability of the battery pack.

The juxtaposition of the simpler EV powertrain against the complex nature of battery pack manufacturing illustrates the evolving challenges in the automotive industry. As manufacturers continue to refine and improve the EV production process, the focus on innovation, safety, and efficiency remains paramount. The transition to electric mobility is reshaping the manufacturing landscape, demanding new skills, technologies, and approaches to meet the unique demands of electric vehicle production.

Innovations in EV Manufacturing

In the realm of electric vehicle (EV) manufacturing, innovation is not merely a buzzword but a fundamental necessity. The drive to make EVs more accessible, efficient, and environmentally friendly propels continuous advancements in manufacturing technologies and processes. A central theme in these innovations is the pursuit of reducing production costs while enhancing the efficiency of the manufacturing process and minimizing its environmental footprint.

Robotic automation has become a cornerstone in the evolution of EV manufacturing, offering unmatched precision and efficiency. These advanced robotic systems are capable of handling complex tasks with high accuracy, from assembling delicate battery components to installing the electric motor, ensuring consistent quality and reducing the likelihood of errors. Their integration into production lines allows for faster assembly times, helping to meet the growing demand for EVs while maintaining high standards of quality.

Moreover, the adoption of innovative materials is revolutionizing the way electric vehicles are built. Lightweight composites and advanced metals are being increasingly used to construct vehicle frames and components. These materials offer the dual benefits of reducing the overall weight of the vehicle, thereby enhancing its energy efficiency and driving range, and improving safety and performance. For example, carbon fiber composites, known for their strength and lightweight properties, are being utilized in body panels and structural components, contributing to the overall reduction in vehicle weight without compromising durability.

The push towards sustainability has significantly influenced manufacturing practices within the EV industry. Manufacturers are increasingly relying on renewable energy sources to power their production facilities, reducing the carbon footprint associated with the manufacturing of electric vehicles. Solar panels and wind turbines are becoming common sights at manufacturing plants, aligning the production of environmentally friendly vehicles with equally sustainable manufacturing processes.

Recycling of materials and the development of closed-loop systems, especially for battery materials, represent another facet of innovation aimed at minimizing waste and environmental impact. By reclaiming valuable metals and materials from used batteries and incorporating them back into the production cycle, manufacturers can reduce the need for new raw materials, lower the environmental impact of mining, and decrease production costs.

These innovations in EV manufacturing showcase the industry's commitment to advancing electric mobility not just through the vehicles themselves but through every stage of their creation. The focus on automation, material science, renewable energy, and recycling underscores a holistic approach to sustainability, aiming to make electric vehicles a cornerstone of a greener, more efficient future in transportation.

Challenges Faced by Manufacturers

In the electric vehicle (EV) manufacturing landscape, the production of lithium-ion batteries stands as a considerable challenge, casting a long shadow over the efforts to mainstream electric mobility. The complexity and high costs associated with battery production are multifaceted issues that manufacturers grapple with, influencing the pace at which EVs can be delivered to an eager market.

The production of lithium-ion batteries, the powerhouse behind EVs, demands significant upfront investments. Building specialized facilities equipped to handle the intricate processes of battery assembly and safety testing is costly, with costs further amplified by the need for high-precision equipment and advanced materials. This financial barrier is a significant hurdle for new entrants to the EV market and represents a considerable portion of the overall cost of electric vehicles.

Beyond the investment in manufacturing facilities, the sourcing and refining of critical materials like lithium, cobalt, and nickel present a web of challenges. These materials are indispensable for the production of high-energy-density batteries, yet their supply is fraught with issues of price volatility and ethical concerns. Prices for these raw materials can fluctuate wildly based on market demand, geopolitical tensions, and regulatory changes, making cost predictions and long-term planning difficult for manufacturers.

The ethical sourcing of these materials adds another layer of complexity. The extraction of lithium, cobalt, and nickel has been linked to environmental degradation, such as water pollution and habitat destruction, and labor violations, particularly in countries with lax environmental regulations and labor laws. Manufacturers are under increasing pressure from consumers and regulatory bodies to ensure their supply chains are not only cost-effective but also ethically and environmentally responsible, necessitating a delicate balance between cost, ethics, and supply chain reliability.

Navigating the rapidly evolving landscape of EV technology and shifting consumer expectations adds to the challenge. The pace of innovation in battery technology is swift, with new advancements around every corner promising longer ranges, shorter charging times, and enhanced safety features. Keeping up with these advancements requires manufacturers to invest heavily in research and development (R&D), pushing the boundaries of current technology to improve battery performance and vehicle capabilities. This investment in R&D is crucial for staying competitive in a market that is increasingly discerning and driven by technological prowess.

Moreover, consumer expectations are continually escalating, driven by a desire for vehicles that offer longer ranges, faster charging, and lower prices. Meeting these expectations while managing the inherent challenges of battery production and ethical sourcing requires a concerted effort from manufacturers to innovate, invest, and adapt to a rapidly changing market landscape.

The challenges faced in the manufacturing of EVs, particularly in battery production, are significant but not insurmountable. Addressing these challenges head-on with innovative solutions, sustainable practices, and a commitment to ethical sourcing is crucial for the continued growth and success of the electric vehicle industry. As manufacturers navigate these hurdles, the future of electric mobility hangs in the balance, promising a cleaner, more sustainable mode of transportation that hinges on the ability to overcome the complex challenges of today.

Supply Chain for Critical Materials

The supply chain for critical materials essential for the production of electric vehicle (EV) batteries represents a labyrinthine global network, entwined with a variety of geopolitical, environmental, and economic considerations. The crux of the issue lies in the geographical concentration of critical mineral reserves necessary for battery production—lithium, cobalt, nickel—within a relatively small number of countries. This geographic concentration not only magnifies the risks of supply disruptions but also exposes manufacturers to the whims of political instability and regulatory changes that can drastically affect availability and pricing.

For instance, the Democratic Republic of Congo (DRC), which harbors the majority of the world's cobalt reserves, is emblematic of the challenges posed by political instability and ethical concerns, including labor practices and environmental protection. Similarly, the dominance of lithium production by countries such as Australia, Chile, and China introduces a layer of complexity in securing a stable and diversified supply chain. The reliance on these concentrated sources exposes manufacturers to significant risks, from sudden changes in export policies to tariffs and trade disputes, each capable of sending ripples through the global supply chain and impacting production timelines and costs.

In response to these vulnerabilities, manufacturers are actively seeking ways to insulate their supply chains from such disruptions and uncertainties. One strategic approach has been to diversify supply sources, not just geographically but also across different materials, to reduce dependence on any single source or country. This strategy involves scouting for new mining operations in politically stable regions with ethical labor practices and environmental standards, thereby spreading risk and ensuring a more resilient supply chain.

Parallel to diversification, there's a concerted push towards investing in alternative battery chemistries that are less reliant on scarce materials. Research and development efforts are increasingly focused on finding viable substitutes for cobalt and lithium or developing new battery technologies that utilize more abundant and easily sourced materials. Such innovations could significantly alter the landscape of battery production, mitigating the environmental and ethical issues associated with mining while also easing supply constraints.

Moreover, manufacturers are engaging in direct partnerships or making strategic investments in mining operations to secure their supply of critical materials. These direct involvement strategies allow for greater oversight and control over the sourcing process, ensuring that materials are extracted and processed in an ethical and environmentally sustainable manner. By forging closer ties with suppliers, manufacturers can achieve greater transparency and reliability in their supply chains, while also promoting sustainable practices in mining operations.

Navigating the complexities of the supply chain for critical battery materials requires a multifaceted approach, blending risk management with ethical considerations and technological innovation. As the EV market continues to expand, the ability of manufacturers to adapt to these challenges and secure a stable, sustainable supply of critical materials will be a determining factor in the pace and scale of the transition to electric mobility.

The Road Ahead for EV Manufacturing

Looking forward, the manufacturing of electric vehicles is set to evolve with advancements in battery technology, automation, and sustainable manufacturing practices. The industry is also likely to see increased localization of supply chains, as manufacturers seek to reduce dependencies, mitigate risks, and comply with regulatory requirements favoring domestic production.

As the EV market continues to grow, the ability of manufacturers to innovate, scale production, and navigate the complex landscape of materials supply will be critical in shaping the future of electric mobility.

Chapter 9: Batteries

The role of batteries in electric vehicles (EVs) transcends mere functionality; they are the linchpin of electric mobility, dictating range, performance, cost, and environmental impact. This section delves into the realm of EV battery technology, exploring the types of batteries used, technological advancements, recycling processes, and sustainability concerns that shape the future of electric transportation.

Types of Batteries Used in EVs

In the landscape of electric vehicles (EVs), the choice of battery technology plays a crucial role in determining the vehicle's range, performance, and overall sustainability. Lithium-ion batteries, with their superior energy density, stand at the forefront of this technology, enabling EVs to achieve longer ranges and more robust performance metrics compared to vehicles powered by other types of batteries. This high energy density means that lithium-ion batteries can store more electricity in a given size, making them particularly suited to the demands of electric mobility where maximizing range while minimizing weight and volume is critical.

The rechargeability of lithium-ion batteries further enhances their appeal for EV use. They can be charged and discharged hundreds of times with relatively minimal loss of capacity, a feature vital for the longevity and practicality of EVs in daily use. This ability to sustain power over numerous charge cycles without significant degradation ensures that electric vehicles remain reliable and efficient over many years of service.

However, the landscape of batteries for EVs is not monolithic. Other battery types, such as nickel-metal hydride (NiMH) and lead-acid, have also found applications within the EV spectrum, though their usage is more niche compared to the dominant lithium-ion technology. NiMH batteries, for example, are lauded for their durability and longer lifespan. These batteries have been a popular choice for hybrid electric vehicles (HEVs), where their robustness and reliability can be leveraged to complement the vehicle's gasoline engine, offering a balance between electric efficiency and the extended range provided by conventional fuels.

Lead-acid batteries, one of the oldest types of rechargeable batteries, are used in EVs as well, albeit in a more constrained role. Due to their lower energy density compared to lithium-ion and NiMH batteries, lead-acid batteries are less suited to serve as the primary power source for electric vehicles. Instead, their application is often relegated to auxiliary functions within EVs, such as powering the electrical systems for lights, dashboard displays, and other vehicle electronics. Despite their limited role in propulsion, lead-acid batteries' low cost, high reliability, and recyclability make them a valuable component of the broader EV ecosystem.

The diversity in battery technology underscores the multifaceted approach to addressing the challenges of electric mobility. Each type of battery brings its own set of strengths and limitations to the table, influencing vehicle design, performance, and user experience. As the EV market continues to evolve, the development of battery technologies remains a dynamic field of innovation, with ongoing research aimed at improving energy density, reducing costs, and enhancing the sustainability profile of these crucial components. The exploration and adoption of various battery types in EVs reflect the industry's commitment to meeting the diverse needs of drivers and paving the way for a more electrified future on the roads.

Advances in Battery Technology

The trajectory of electric vehicle (EV) battery technology is marked by relentless pursuit of innovation, driven by the imperative to enhance performance, affordability, and environmental sustainability. At the heart of this quest is the continuous advancement in battery technology, which holds the key to unlocking the full potential of EVs. Researchers and engineers are dedicated to improving the fundamental aspects of battery performance — energy density, longevity, charging speed, and cost-effectiveness — each of which plays a critical role in the practicality and appeal of electric vehicles.

Energy density, a measure of how much power can be stored in a given volume, is a primary focus of battery innovation. Higher energy densities enable longer driving ranges without the need for excessively large or heavy battery packs, directly addressing one of the most significant consumer concerns about EVs. Advances in battery chemistry and cell design are progressively increasing the amount of energy that batteries can store, thereby extending the distances EVs can travel on a single charge.

Extending battery life is another crucial area of development. The longer a battery can maintain its capacity over time and through repeated charging cycles, the longer the lifespan of the EV and the lower the total cost of ownership. Innovations in battery management systems (BMS) and electrode materials are improving battery durability, ensuring that EV batteries retain their charge capacity and performance for increasingly extended periods.

Reducing charging times is essential for making EVs as convenient as their gasoline counterparts. Rapid charging technologies are evolving, aiming to reduce the time it takes to recharge an EV to minutes rather than hours. These advancements, coupled with the expansion of fast-charging infrastructure, are making

EVs more suitable for a wider range of travel needs, including long-distance journeys.

Cost reduction remains a pivotal challenge and area of innovation in battery technology. Lowering the cost of batteries directly affects the overall price of EVs, making them more accessible to a broader audience. Economies of scale, improvements in manufacturing processes, and the development of less expensive battery materials are contributing to the ongoing decrease in battery costs.

Among the most promising innovations in the field are solid-state batteries, which replace the liquid electrolyte found in conventional lithium-ion batteries with a solid alternative. Solid-state batteries boast potential advantages, including higher energy densities, enhanced safety due to reduced risk of leakage and thermal runaway, and longer lifespans. These batteries represent a significant leap forward in battery technology, with the potential to address many of the limitations of current lithium-ion batteries.

Further on the horizon are lithium-sulfur (Li-S) and lithium-air (Li-air) batteries. These technologies offer even greater energy densities and the promise of further reducing the environmental impact of batteries. Li-S batteries, with their higher theoretical energy density, could provide much longer ranges for EVs, while Li-air batteries hold the potential for unprecedented energy storage capacity. Both technologies are still in the research phase, with challenges to overcome in terms of durability, efficiency, and cost.

The continuous advancement in battery technology is not just about enhancing the technical specifications of EVs. It is also about addressing the broader implications of electric mobility, from environmental sustainability to the global economy. As these innovations progress from laboratory breakthroughs to commercial applications, they pave the way for a future where electric vehicles are the standard, not the exception, in transportation.

Recycling and Sustainability Issues

As electric vehicles (EVs) continue to proliferate, the spotlight intensifies on the sustainability of EV batteries, especially regarding their end-of-life management and overall lifecycle environmental impact. The imperative to establish efficient and environmentally sound recycling processes for these batteries is becoming increasingly critical. Effective recycling serves multiple purposes: it significantly mitigates the environmental burden by curtailing the demand for virgin raw materials, and it addresses potential disposal concerns related to the hazardous chemicals contained in batteries.

The standard recycling process for EV batteries encompasses several steps, beginning with the disassembly of battery packs to separate out the individual cells or modules. Following disassembly, these components are typically shredded, a process that breaks them down into smaller pieces, allowing for the subsequent extraction of valuable materials such as lithium, cobalt, and nickel. These materials are essential not only for the production of new batteries but also for various applications in the electronics industry, making their recovery both an environmental and economic priority.

Despite the established recycling protocols, the sector faces challenges in enhancing the efficiency of material recovery and ensuring the economic viability of the recycling process itself. Current methodologies for extracting valuable metals from shredded battery components vary in efficiency and often involve energy-intensive procedures that can diminish the environmental benefits of recycling. Moreover, the economic aspect of recycling—balancing the cost of recycling processes against the value of recovered materials—remains a critical area for improvement. As such, ongoing innovation in recycling technology is crucial, aiming to bolster material recovery rates while simultaneously reducing the energy footprint of the recycling process.

In addition to recycling, the sustainability of battery production from inception to disposal raises significant concerns. The environmental and ethical implications of mining the raw materials required for battery manufacture, such as lithium and cobalt, are well-documented, encompassing habitat destruction, water pollution, and labor rights issues. In response, there is a concerted push towards developing more sustainable mining practices that minimize environmental damage and ensure fair labor conditions. Concurrently, the shift towards utilizing renewable energy sources in battery manufacturing processes is gaining momentum, aiming to decrease the carbon footprint associated with the production of EV batteries.

Exploration into alternative battery chemistries that rely on more abundantly available and less environmentally problematic materials is another facet of the industry's effort to enhance sustainability. By reducing reliance on scarce or ethically contentious materials, these new chemistries promise to alleviate some of the environmental and social issues currently associated with battery production.

The trajectory towards sustainable EV battery production and recycling is complex, necessitating a multi-pronged approach that addresses the environmental, economic, and social dimensions of sustainability. As the EV market grows, the development of advanced recycling technologies, the adoption of greener production practices, and the exploration of innovative battery materials are pivotal in ensuring that electric mobility contributes positively to the global sustainability agenda. This holistic approach to sustainability is fundamental to the long-term success and acceptance of electric vehicles, ensuring that the transition to electric mobility is truly beneficial for the planet and its inhabitants.

The Road Ahead

The future of EV battery technology is a convergence of innovation, sustainability, and economics. As advances in battery chemistry and design continue to unfold, the focus remains on enhancing performance and reducing the environmental impact of batteries throughout their lifecycle. The evolution of battery technology, coupled with improvements in recycling and sustainability practices, will play a pivotal role in the transition to a more sustainable and electrified transportation future.

Chapter 10: Real Impact on the Environment

The environmental impact of electric vehicles (EVs) is a subject of extensive debate and analysis, focusing on the entire lifecycle from production through to disposal. This objective examination considers not only the direct emissions—or lack thereof—from the vehicles themselves but also the broader implications of their production, energy consumption, and end-of-life processes. Additionally, comparing the carbon footprint of EVs with that of internal combustion engine (ICE) vehicles offers insights into the potential benefits and challenges associated with the transition to electric mobility.

Lifecycle Environmental Impact

The lifecycle environmental impact of electric vehicles (EVs) presents a complex narrative that stretches from the initial extraction of raw materials to the ultimate disposal or recycling of the vehicle. This comprehensive analysis sheds light on both the environmental advantages and the challenges associated with EVs, offering a nuanced perspective on their role in the transition towards sustainable mobility.

The journey of an EV begins long before it hits the road, starting with the extraction of raw materials needed for its construction. The production of EV batteries, a cornerstone of electric mobility, requires significant quantities of lithium, cobalt, nickel, and other minerals. The mining of these materials is often fraught with environmental concerns. For instance, lithium extraction, predominantly through the process of evaporation of large brine pools, can lead to water scarcity and ecological disruption in arid regions. Cobalt mining, particularly in the Democratic Republic of Congo, has been associated with severe human rights abuses and ecological harm. Nickel mining, required for battery cathodes, poses risks of water pollution and forest degradation. These issues underscore the environmental and ethical complexities of sourcing materials for EV batteries, highlighting the need for more sustainable mining practices and the exploration of alternative materials.

The manufacturing phase further contributes to the environmental footprint of EVs. Assembling the vehicle and its components, especially the battery pack, is energy-intensive. When the energy used in these processes comes from fossil fuels, it significantly increases the vehicle's carbon footprint even before it leaves the factory. This stage underscores the importance of transitioning to cleaner energy sources within the manufacturing sector and enhancing the energy efficiency of production processes. Advances in manufacturing technology, such as the use of renewable energy and more efficient production lines, are crucial in minimizing the carbon footprint associated with EV manufacturing.

Once on the road, EVs exhibit a clear environmental advantage over internal combustion engine (ICE) vehicles by producing zero tailpipe emissions. This characteristic is particularly beneficial in urban environments, where vehicular emissions are a major contributor to air pollution and associated health problems. The shift to EVs can significantly improve urban air quality, contributing to better respiratory health and overall public welfare.

However, the environmental impact of operating an EV is intrinsically linked to the source of the electricity used for charging. In regions where the grid is predominantly powered by renewable energy sources, such as hydroelectric, wind, or solar power, EVs can operate with minimal environmental impact, offering a substantial reduction in lifecycle emissions. Conversely, in areas where the grid relies heavily on coal or other fossil fuels, the benefits of switching to EVs are diminished, as the carbon emissions associated with electricity generation offset the advantages of eliminating tailpipe emissions.

The end-of-life management of EVs, particularly the recycling of batteries, is a critical aspect of their lifecycle environmental impact. Effective recycling can mitigate some of the environmental concerns associated with the production of EVs by reducing the demand for new raw materials and minimizing waste. Current recycling processes focus on recovering valuable materials from batteries, but improving the efficiency of these processes and extending them to other vehicle components is essential. The development of sustainable recycling technologies and the establishment of comprehensive recycling infrastructure are vital steps in ensuring that the environmental benefits of EVs are fully realized throughout their lifecycle.

In conclusion, the lifecycle analysis of EVs reveals a multifaceted environmental impact that encompasses both significant benefits and notable challenges. While EVs offer the promise of cleaner transportation with zero tailpipe emissions, realizing their full environmental potential requires addressing the impacts associated with raw material extraction, manufacturing, and end-of-life management. As the energy grid becomes cleaner and advancements in battery technology and recycling continue, the lifecycle environmental impact of EVs is expected to improve, solidifying their role in a sustainable transportation future.

Comparison with ICE Vehicles

The comparison between electric vehicles (EVs) and internal combustion engine (ICE) vehicles regarding their environmental impact involves a comprehensive analysis of total emissions across their respective lifecycles. This analysis extends from the manufacturing process, through the operational life, to the disposal or recycling of the vehicle. While EVs present higher initial emissions primarily due to the energy-intensive production of batteries, they offer significantly reduced operational emissions, attributed to their zero-emission nature during use. This contrast highlights the evolving dynamics of automotive environmental impacts, especially when considering the source of electricity for charging EVs and the advancements in technology and regulatory frameworks aimed at reducing emissions.

ICE vehicles, reliant on gasoline or diesel, contribute to CO2 and other pollutant emissions with every mile driven, directly impacting air quality and public health. The combustion process inherent to these vehicles' operation makes their environmental footprint during use substantially higher than that of EVs. Although improvements in fuel efficiency and cleaner fuels have reduced the per-mile emissions of ICE vehicles, they cannot match the lower operational emissions profile of electric vehicles, especially as the electricity grid transitions to renewable energy sources.

The lifecycle carbon footprint analysis of EVs versus ICE vehicles underscores a critical factor: the greening of the electricity grid plays a pivotal role in amplifying the environmental advantages of EVs. As renewable energy sources like wind, solar, and hydro increasingly replace fossil fuels in electricity generation, the operational emissions of EVs continue to decrease. Consequently, the gap in lifecycle emissions between EVs and ICE vehicles widens, affirming the long-term environmental benefits of electric mobility.

Moreover, advancements in battery technology and manufacturing processes are progressively reducing the upfront emissions associated with EVs. Innovations in battery design, increased manufacturing efficiency, and the development of recycling technologies for battery materials are contributing to a decrease in the environmental impact of the production stage. Regulatory measures aimed at reducing the carbon intensity of both the automotive manufacturing sector and the power generation industry further bolster the environmental case for EVs.

In the broader context of alternative fuel vehicles, the comparison between EVs and hydrogen fuel cell vehicles (FCVs) adds another dimension to the discussion. Like EVs, FCVs offer the promise of zero tailpipe emissions, with water vapor being the only byproduct of their operation. However, the environmental impact of hydrogen production, which is energy-intensive, especially when derived from fossil fuels, can offset some of the benefits of FCVs. Additionally, the current lack of hydrogen refueling infrastructure and the challenges associated with hydrogen storage and transportation present hurdles to the widespread adoption of FCVs.

Comparatively, EVs benefit from a more established charging infrastructure and the ongoing rapid expansion of renewable energy sources for electricity generation, positioning them favorably in the context of immediate and scalable solutions for reducing transportation emissions. As the global energy landscape evolves towards sustainability, the environmental credentials of EVs are expected to strengthen, underscoring their role in the transition to a low-carbon future.

In conclusion, while both EVs and ICE vehicles present distinct profiles in terms of lifecycle emissions, the trajectory of technological and regulatory developments suggests a growing environmental advantage for electric mobility. The comparison, enriched by the consideration of hydrogen fuel cell vehicles, illuminates the diverse pathways towards sustainable transportation, highlighting the critical role of energy sources and technological innovation in shaping the environmental impact of our mobility choices.

Impact of Electricity Generation

The environmental impact of electric vehicles (EVs) is intricately linked to the method by which electricity is generated to charge them. The composition of the energy mix used for electricity generation plays a pivotal role in determining the overall carbon footprint of EVs and, consequently, their environmental benefits. As countries worldwide strive to decarbonize their energy systems, transitioning towards renewable energy sources such as wind, solar, and hydroelectricity becomes imperative for maximizing the environmental advantages of electric mobility.

The increasing adoption of EVs presents both opportunities and challenges in the context of electricity generation. On one hand, the widespread adoption of EVs can serve as a catalyst for accelerating the transition to renewable energy sources. The growing demand for electricity from EV charging can incentivize investments in renewable energy infrastructure, driving down costs and expanding the share of clean energy in the grid. This virtuous cycle not only reduces greenhouse gas emissions from transportation but also contributes to overall emissions reductions by displacing fossil fuel-based electricity generation.

However, the transition to electric mobility also poses challenges for electricity generation infrastructure, particularly in regions where renewable energy penetration is low or where the grid is heavily reliant on fossil fuels. The increased demand for electricity from EVs may necessitate the construction of additional power plants, which, if not powered by renewable sources, could exacerbate environmental concerns related to air pollution and carbon emissions. In countries like China, where coal-fired power stations still constitute a significant portion of the energy mix, the environmental benefits of EVs may be diminished by the emissions associated with electricity generation.

Addressing these challenges requires a multifaceted approach that prioritizes the decarbonization of the electricity grid while simultaneously promoting energy efficiency and conservation measures. Policy interventions, such as renewable energy targets, carbon pricing mechanisms, and incentives for clean energy investment, are essential for accelerating the transition to a low-carbon energy future. Additionally, efforts to improve grid flexibility, energy storage technologies, and demand-side management strategies can help optimize the integration of renewable energy sources and mitigate the environmental impacts of increased electricity demand from EVs.

Furthermore, ensuring equitable access to clean and affordable electricity is crucial for maximizing the environmental and social benefits of electric mobility.

In many regions, vulnerable communities disproportionately bear the burden of air pollution and climate change impacts, highlighting the importance of implementing policies that prioritize environmental justice and promote inclusive and sustainable development.

As technology advances and policy frameworks evolve, the environmental benefits of electric mobility are expected to grow, offering a promising pathway to reducing transportation-related emissions and combating climate change. However, realizing the full potential of EVs in fostering a sustainable transport system requires concerted efforts to decarbonize the electricity grid, enhance energy efficiency, and promote equitable access to clean energy. By embracing a holistic approach to electric mobility that integrates environmental, social, and economic considerations, countries can harness the transformative power of EVs to create a cleaner, healthier, and more sustainable future for all.

Chapter 11: Charging and Infrastructure

The development of robust charging infrastructure is critical for the widespread adoption and success of electric vehicles (EVs). This section explores the various types of charging stations, efforts to expand infrastructure, and the evolving landscape of charging technology, including advancements in wireless charging and fast charging networks.

Types of Charging Stations

Charging infrastructure for electric vehicles (EVs) encompasses a variety of charging stations, each tailored to different use cases and charging requirements. These charging stations offer distinct levels of power output and charging speeds, catering to the diverse needs of EV drivers.

Level 1 charging represents the most basic form of EV charging, utilizing a standard household outlet operating at 120 volts. While convenient and accessible, level 1 charging provides the slowest charging rate among the different options. This charging method is typically suitable for overnight charging at home or in situations where a quick top-up is sufficient, such as during emergencies or short stops.

In contrast, level 2 charging significantly boosts charging speeds by utilizing a 240-volt circuit. Level 2 charging stations are commonly installed at homes, workplaces, and public locations, offering faster charging rates than level 1 chargers. This level of charging is well-suited for daily use, allowing EV drivers to conveniently replenish their vehicle's battery while parked for extended periods, such as during work hours or shopping trips.

Level 3 charging, also known as DC fast charging, represents the fastest charging option available for EVs. These charging stations deliver a significant charge in a short amount of time, making them indispensable for long-distance travel and expedited charging needs. Level 3 chargers utilize direct current (DC) power, bypassing the vehicle's onboard charger to deliver electricity directly to the battery at high rates. These chargers are typically found along highways, major travel routes, and key transportation corridors, enabling EV drivers to quickly recharge their vehicles during extended journeys.

Each type of charging station serves a specific purpose within the broader charging infrastructure ecosystem, catering to the diverse needs and preferences of EV drivers. Level 1 chargers offer convenience and accessibility for everyday charging needs, while level 2 chargers provide faster charging rates for extended parking durations. Level 3 fast chargers offer rapid charging capabilities, facilitating long-distance travel and minimizing charging downtime for drivers on the go.

As the EV market continues to evolve and grow, the expansion and diversification of charging infrastructure remain essential to support the increasing adoption of electric mobility. By offering a range of charging options tailored to different scenarios and use cases, charging stations play a crucial role in enabling the widespread adoption of electric vehicles and promoting sustainable transportation solutions for the future.

Infrastructure Development Efforts

Efforts to expand charging infrastructure for electric vehicles (EVs) are gaining momentum on a global scale, fueled by a combination of government incentives, private investments, and collaborative initiatives. Governments around the world are implementing policies and programs aimed at incentivizing the installation and expansion of charging stations to support the growing EV market. These initiatives include a range of measures such as grants, tax credits, and regulatory mandates designed to encourage investment in charging infrastructure and accelerate its deployment.

Private companies play a significant role in the expansion of charging networks, with utilities and charging network operators investing heavily to meet the increasing demand for EV charging services. These companies recognize the potential for growth in the EV market and are actively expanding their charging infrastructure to capture market share and provide convenient charging options for EV owners. By investing in the development of charging networks, private entities contribute to the accessibility and reliability of EV charging, driving further adoption of electric vehicles.

Collaborative efforts between stakeholders are also driving the expansion of charging infrastructure. Automakers, utilities, municipalities, and other key players are joining forces to streamline infrastructure deployment and ensure interoperability between charging networks. By coordinating their efforts and sharing resources, these stakeholders can overcome challenges such as permitting, zoning, and grid integration, accelerating the rollout of charging infrastructure and improving the overall charging experience for EV drivers.

The collective efforts of governments, private companies, and stakeholders are instrumental in expanding charging infrastructure and supporting the transition to electric mobility. By creating an enabling environment for investment and collaboration, policymakers can stimulate innovation and drive the development of robust and accessible charging networks. Private sector investments complement government incentives by expanding the reach and capacity of charging infrastructure, while collaborative initiatives foster coordination and cooperation among key stakeholders, ensuring the efficient deployment and operation of charging networks.

As the EV market continues to grow and evolve, ongoing efforts to expand charging infrastructure will be essential to meet the increasing demand for EV charging services and support the widespread adoption of electric vehicles. By leveraging government incentives, private investments, and collaborative partnerships, stakeholders can build a comprehensive and reliable charging infrastructure that enables convenient and accessible charging for all EV drivers, driving the transition to a cleaner, more sustainable transportation future.

Future of Charging Technology

The future of charging technology for electric vehicles (EVs) is marked by continuous innovation and advancement, aimed at enhancing charging speeds, convenience, and efficiency. Ongoing research and development efforts are focused on exploring new technologies and improving existing solutions to address the evolving needs of EV drivers and support the widespread adoption of electric mobility.

One of the most promising developments in charging technology is wireless charging, which eliminates the need for physical cables and connectors. Wireless charging offers the potential for effortless charging experiences at home, in parking lots, and even on roads, enabling EV drivers to charge their vehicles conveniently without the hassle of plugging in. This technology has the potential to revolutionize the EV charging experience, making it more seamless and user-friendly, while also reducing wear and tear on charging connectors and cables.

Fast charging networks represent another key area of innovation in charging technology. These networks are equipped with high-power chargers capable of delivering ultra-rapid charging speeds, significantly reducing charging times and enhancing the practicality of EVs for long-distance travel. By deploying fast charging infrastructure along major highways and travel routes, stakeholders aim to address range anxiety and promote the adoption of electric mobility by offering drivers convenient access to fast and reliable charging facilities.

Advancements in battery technology are also driving improvements in charging speeds and efficiency. Technologies such as solid-state batteries, which offer higher energy density and faster charging rates compared to traditional lithium-ion batteries, hold the potential to enable faster and more efficient charging experiences without compromising battery longevity or safety. These advancements in battery technology are essential for enhancing the overall performance and usability of electric vehicles, making them more competitive with conventional vehicles in terms of range, charging time, and convenience.

Smart charging solutions are another area of innovation in charging technology, aimed at optimizing charging schedules and minimizing the impact of EV charging on electricity grids. By integrating renewable energy sources such as solar and wind power, smart charging systems can prioritize charging during periods of low demand or high renewable energy availability, reducing reliance on fossil fuels and promoting sustainable charging practices. Additionally, smart charging solutions can enable vehicle-to-grid (V2G) capabilities, allowing EVs to serve as grid storage devices and provide valuable grid services, such as demand response and frequency regulation.

Overall, the future of EV charging holds great promise, with advancements in technology driving improvements in charging speeds, convenience, and sustainability. By investing in diverse charging solutions, fostering collaboration between stakeholders, and embracing innovative technologies, the EV charging ecosystem can continue to evolve and expand, overcoming barriers to adoption and ensuring the seamless integration of electric mobility into daily life. As the EV market continues to grow, the expansion and enhancement of charging infrastructure will play a crucial role in supporting the transition to a cleaner, more sustainable transportation future.

Chapter 12: Legislation

Legislation plays a pivotal role in shaping the adoption and development of electric vehicles (EVs) on a global scale. This section explores the diverse landscape of laws and regulations that influence the EV market, encompassing emissions regulations, safety standards, and incentives for manufacturers and consumers.

Emissions Regulations

The imperative to mitigate climate change and reduce greenhouse gas emissions has propelled governments worldwide to enact stringent emissions regulations, particularly targeting the automotive sector. Recognizing the significant contribution of transportation to overall emissions, policymakers have prioritized measures to incentivize the adoption of low- and zero-emission vehicles, with electric cars emerging as a prominent solution in the transition to cleaner transportation.

Emissions regulations typically impose targets for vehicle manufacturers to reduce the average emissions of their fleets, incentivizing the production and sale of electric vehicles (EVs) to achieve compliance. These targets are often framed in terms of grams of CO_2 emitted per kilometer driven, with manufacturers required to meet increasingly stringent standards over time. By imposing such regulations, policymakers aim to drive technological innovation and investment in cleaner vehicle technologies, ultimately accelerating the shift away from fossil fuel-powered vehicles towards more sustainable alternatives.

In addition to setting emissions targets, governments may also impose penalties for non-compliance, providing further incentive for manufacturers to prioritize the production and sale of EVs. These penalties may take the form of fines or other regulatory measures, creating financial disincentives for manufacturers that fail to meet emissions standards. By enforcing penalties for non-compliance, policymakers send a clear signal to the automotive industry of the importance of reducing emissions and transitioning to cleaner vehicle technologies.

Furthermore, emissions regulations are often part of broader policy frameworks aimed at addressing climate change and promoting sustainable transportation. These frameworks may include a combination of regulatory measures, financial incentives, and investment in infrastructure to support the transition to electric mobility. By aligning emissions regulations with broader policy objectives, governments can create a supportive environment for the adoption of EVs and other clean transportation technologies.

Overall, emissions regulations play a crucial role in driving the transition to electric mobility and reducing greenhouse gas emissions from the transportation sector. By setting targets for emissions reduction, imposing penalties for non-compliance, and aligning regulations with broader policy objectives, governments can incentivize the production and adoption of EVs, paving the way for a more sustainable transportation future.

Safety Standards

The paramount importance of ensuring the safety of electric vehicles (EVs) cannot be overstated in fostering their widespread adoption and acceptance among consumers. Legislation governing vehicle safety standards plays a pivotal role in establishing requirements for crashworthiness, occupant protection, pedestrian safety, and other essential criteria.

These safety standards are meticulously crafted to ensure that EVs meet rigorous safety requirements before being made available to consumers. EV manufacturers are mandated to adhere to these standards to obtain regulatory approval and certification for their vehicles. Compliance with safety standards is non-negotiable, serving as a crucial aspect of regulatory oversight to safeguard the well-being of vehicle occupants, pedestrians, and other road users.

Safety standards for EVs encompass various aspects, including structural integrity, crash mitigation systems, restraint systems (such as seat belts and airbags), braking performance, and electronic stability control. Additionally, considerations are made to address the unique characteristics of EVs, such as battery safety and electrical system integrity, to mitigate the risk of fire or electrical hazards.

In addition to regulatory requirements, governments may incentivize EV manufacturers to prioritize safety in their designs by offering incentives or subsidies contingent upon vehicles meeting specified safety standards. By aligning financial incentives with safety compliance, policymakers further underscore the importance of safety in the development and production of EVs.

Ultimately, adherence to safety standards is essential not only for regulatory compliance but also for maintaining consumer confidence and trust in EVs. By prioritizing safety in design and engineering, manufacturers can enhance the overall safety performance of EVs, contributing to their widespread adoption and integration into the global transportation landscape. Through a combination of stringent safety regulations, regulatory oversight, and incentivization, governments play a vital role in promoting the safety of EVs and fostering consumer confidence in this transformative mode of transportation.

Global and Country-Specific Legislation

While there are overarching trends in electric vehicle (EV) legislation, specific laws and regulations vary significantly by country and region, reflecting diverse priorities, challenges, and opportunities. Let's delve into the regulatory landscapes of key countries and regions, including the United States, United Kingdom, European Union, China, and India, to understand the nuanced approaches to EV legislation and their impact on electric mobility.

United States:

In the United States, electric vehicle (EV) legislation is shaped by a combination of federal, state, and local regulations, reflecting the country's federalist system of governance. At the federal level, the government provides a range of tax credits and incentives to encourage EV adoption and support the development of EV technology. These incentives include a federal tax credit of up to $7,500 for the purchase of qualifying

EVs, aimed at offsetting the higher upfront costs associated with electric vehicles. Additionally, federal agencies allocate funding for research and development in EV technology, supporting advancements in battery technology, charging infrastructure, and electric drivetrains.

States play a significant role in shaping EV legislation, with many adopting their own incentives and regulations to promote electric mobility. California, in particular, has been a trailblazer in this regard, implementing ambitious Zero Emission Vehicle (ZEV) mandates to drive EV adoption. These mandates require automakers to sell a certain percentage of zero-emission vehicles, including battery electric vehicles (BEVs) and fuel cell electric vehicles (FCEVs), in the state's vehicle market. By setting targets for automakers to sell a minimum number of EVs, California aims to reduce greenhouse gas emissions and improve air quality.

In addition to ZEV mandates, various states across the country offer additional incentives to promote EV adoption. These incentives may include rebates or tax credits for EV purchases, reduced registration fees, access to high-occupancy vehicle (HOV) lanes, and grants for the installation of charging infrastructure. By providing financial incentives and other benefits to EV owners, states seek to stimulate consumer demand for electric vehicles and accelerate the transition away from traditional internal combustion engine vehicles.

Furthermore, local governments, such as cities and municipalities, also play a role in promoting electric mobility through zoning regulations, building codes, and incentives for EV charging infrastructure deployment. Some localities offer incentives for businesses to install charging stations or require new developments to include EV charging infrastructure as part of their construction plans. These efforts contribute to the expansion of charging infrastructure and facilitate EV adoption in urban areas.

Overall, the United States has adopted a multifaceted approach to promoting electric mobility, with federal, state, and local governments working together to incentivize EV adoption and support the growth of the EV market. By providing financial incentives, implementing regulations, and investing in infrastructure, policymakers aim to reduce emissions, improve air quality, and foster innovation in the transportation sector.

United Kingdom:
In the United Kingdom (UK), the government has embarked on an ambitious journey towards phasing out internal combustion engine vehicles, aligning its policies with broader climate and environmental goals. Central to this effort is the commitment to ban the sale of new petrol and diesel cars by 2030, marking a significant milestone in the transition to electric mobility.

Legislation in the UK reflects a comprehensive strategy aimed at incentivizing electric vehicle (EV) adoption and supporting the necessary infrastructure development. One of the key components of this strategy is the provision of financial incentives for EV purchases. The government offers grants and subsidies to reduce the upfront costs of EVs, making them more accessible and affordable to consumers. These incentives serve to stimulate demand for electric vehicles and encourage consumers to make the switch from traditional internal combustion engine vehicles.

Additionally, the UK government provides tax exemptions and other financial benefits to EV owners, further enhancing the attractiveness of electric mobility. These incentives may include reduced vehicle excise duty (road tax), exemption from congestion charges in urban areas, and preferential parking privileges for EVs. By offering these incentives, policymakers aim to create a favorable environment for EV adoption and accelerate the transition away from fossil fuel-powered vehicles.

Furthermore, the UK government recognizes the importance of investing in charging infrastructure to support the growing fleet of electric vehicles. Legislation includes funding initiatives to expand and enhance the country's charging network, ensuring that EV drivers have convenient access to charging facilities wherever they go. This investment in charging infrastructure is critical for alleviating range anxiety and addressing one of the key barriers to EV adoption.

In addition to infrastructure investment, the UK government is committed to promoting innovation in

EV technology. This includes funding research and development initiatives to drive advancements in battery technology, electric drivetrains, and charging technology. By supporting innovation in the EV sector, policymakers aim to enhance the performance, affordability, and sustainability of electric vehicles, further driving their adoption and market penetration.

Overall, the UK's approach to EV legislation reflects a holistic strategy aimed at accelerating the transition to electric mobility. By setting ambitious targets, providing financial incentives, investing in infrastructure, and promoting innovation, the government seeks to reduce emissions, improve air quality, and lead the way towards a more sustainable transportation future.

European Union (EU):
The European Union (EU) has taken proactive steps to address climate change and promote the adoption of electric vehicles (EVs) through a combination of stringent emissions regulations, incentives, and directives aimed at driving the transition to cleaner transportation.

Central to the EU's efforts is the establishment of stringent emissions regulations, which include fleet-wide CO2 emission targets for automakers. These targets set limits on the average CO2 emissions produced by vehicles sold within the EU, incentivizing automakers to produce and sell low-emission and zero-emission vehicles, such as electric cars. By imposing these targets, the EU aims to reduce greenhouse gas emissions from the transportation sector and mitigate climate change.

In addition to emissions regulations, countries within the EU offer a variety of incentives and regulations to promote EV adoption. These incentives may include purchase subsidies, tax credits, and rebates aimed at reducing the upfront costs of EVs and making them more affordable to consumers. Furthermore, some EU countries provide exemptions from road taxes, congestion charges, and other levies for EV owners, incentivizing the switch to electric mobility.

The EU's Clean Vehicles Directive is another important legislative tool aimed at promoting the adoption of clean vehicles, including EVs. The directive encourages public procurement of clean vehicles by government agencies and public transportation fleets, thereby increasing the demand for EVs and driving market growth. By prioritizing the procurement of clean vehicles, the EU aims to lead by example and accelerate the transition to a greener transportation system.

Overall, the EU's approach to EV legislation reflects a comprehensive strategy aimed at reducing emissions, improving air quality, and fostering innovation in the automotive sector. By combining stringent emissions regulations with incentives and directives to promote EV adoption, the EU seeks to create a supportive environment for electric mobility and pave the way for a more sustainable transportation future in Europe.

China:
China's position as the world's largest electric vehicle (EV) market is a testament to its robust government policies and incentives aimed at accelerating the adoption of electric mobility and addressing environmental challenges. At the forefront of China's EV strategy are aggressive government policies and

subsidies designed to stimulate both demand and supply within the EV market.

The Chinese government offers generous subsidies for EV purchases, significantly reducing the upfront costs for consumers and making electric vehicles more financially accessible. These subsidies have played a crucial role in driving consumer demand for EVs and incentivizing the transition away from traditional internal combustion engine vehicles.

In addition to subsidies for consumers, China provides incentives for manufacturers to produce EVs, fostering innovation and investment in the domestic EV industry. These incentives may include financial support, tax breaks, and other preferential policies aimed at encouraging automakers to develop and produce electric vehicles. By incentivizing EV production, the Chinese government aims to stimulate economic growth, create jobs, and establish China as a global leader in electric mobility.

Furthermore, China has implemented Zero Emission Vehicle (ZEV) mandates in certain regions, requiring automakers to sell a minimum percentage of EVs in their vehicle fleets. These mandates serve to increase the availability and variety of electric vehicles in the market, further driving consumer adoption and market growth.

To support the growing EV fleet, the Chinese government is investing heavily in charging infrastructure, recognizing the importance of reliable and accessible charging facilities in facilitating the widespread adoption of electric vehicles. This investment includes the deployment of public charging stations in urban areas, along highways, and in key transportation hubs, ensuring that EV owners have convenient access to charging facilities wherever they go.

Overall, China's success as a global leader in electric mobility can be attributed to its comprehensive approach to EV legislation, which combines generous subsidies, incentives for manufacturers, ZEV mandates, and investments in charging infrastructure. By leveraging these policies and incentives, China aims to reduce emissions, improve air quality, and drive innovation in the automotive sector, positioning itself for a sustainable and prosperous future in electric mobility.

India:

In India, the landscape of electric vehicle (EV) legislation is evolving rapidly as the government embarks on ambitious plans to promote electric mobility and address environmental challenges. While still in its nascent stages, India's EV legislation reflects a growing recognition of the importance of transitioning to cleaner transportation solutions and reducing dependence on fossil fuels.

The Indian government has announced several initiatives aimed at incentivizing the adoption and production of electric vehicles. These initiatives include incentives for EV manufacturing, such as subsidies, tax breaks, and other financial incentives designed to stimulate investment and innovation in the domestic EV industry. By offering incentives for manufacturers, the government aims to create a favorable environment for the development and production of electric vehicles, driving growth and competitiveness in the sector.

In addition to incentives for manufacturing, the Indian government provides research and development grants to support innovation and technological advancements in the EV sector. These grants facilitate collaboration between industry, academia, and research institutions, driving innovation and enhancing the performance, efficiency, and affordability of electric vehicles.

Furthermore, the Indian government offers tax incentives for EV buyers, reducing the upfront costs of purchasing electric vehicles and making them more financially attractive to consumers. These incentives may include exemptions from sales tax, registration fees, and road tax, as well as subsidies for EV purchases. By providing financial incentives for consumers, the government aims to stimulate demand for electric vehicles and accelerate their adoption across the country.

To support the widespread adoption of electric vehicles, the Indian government has set ambitious targets for EV penetration, aiming to achieve 30% EV penetration by 2030. To realize this vision, the government has proposed various measures to incentivize EV adoption, including subsidies for EV purchases, investments in charging infrastructure development, and other supportive policies and initiatives.

Overall, India's approach to EV legislation reflects a commitment to promoting electric mobility and transitioning towards a more sustainable transportation system. By offering incentives for EV manufacturing and research, providing tax incentives for consumers, and setting ambitious targets for EV penetration, the Indian government aims to accelerate the adoption of electric vehicles and reduce emissions, improving air quality and mitigating the impacts of climate change.

In summary, legislation plays a pivotal role in shaping the trajectory of electric mobility, with each country adopting a unique approach based on its specific priorities and circumstances. By implementing supportive policies and regulations, governments can accelerate the transition to electric vehicles, reduce emissions, and pave the way for a more sustainable transportation future on a global scale.

Chapter 13: Your Guide to Buying an Electric Vehicle in 2024

Understanding EVs

You've made the bold decision to embrace the future of transportation by considering an electric vehicle (EV) for your next ride. But what exactly is an electric vehicle, and how does it differ from the conventional cars you're used to?

Electric vehicles, or EVs, represent a revolutionary shift in automotive technology. Unlike traditional internal combustion engine vehicles, which rely on gasoline or diesel to power their engines, EVs are propelled by electric motors fueled by rechargeable batteries. These batteries store electricity, which is then used to power the vehicle's motor, resulting in smooth and silent acceleration without the need for burning fossil fuels.

The transition to electric mobility offers numerous advantages over traditional gasoline-powered cars. One of the most significant benefits is environmental sustainability. EVs produce zero tailpipe emissions, meaning they don't release harmful pollutants such as carbon dioxide (CO_2), nitrogen oxides (NO_x), and particulate matter into the air during operation. This reduction in emissions contributes to cleaner air quality and helps mitigate the adverse effects of climate change caused by greenhouse gas emissions.

Moreover, electric vehicles are inherently more energy-efficient than their internal combustion engine counterparts. Electric motors are capable of converting a higher percentage of the energy stored in batteries into mechanical power, resulting in greater efficiency and reduced energy waste compared to the complex combustion process in traditional engines. This efficiency translates into lower fuel costs and decreased reliance on finite fossil fuels, ultimately leading to long-term savings for EV owners.

Additionally, electric vehicles offer a quieter and smoother driving experience thanks to the absence of noisy engine combustion and vibrations associated with internal combustion engines. The instant torque delivery of electric motors provides responsive acceleration, making EVs feel nimble and agile on the road.

As you delve deeper into the world of electric mobility, you'll discover a wide range of EV models available, from compact hatchbacks to luxurious SUVs and high-performance sports cars. Each EV offers its own unique blend of features, driving range, and charging capabilities, allowing you to find the perfect fit for your lifestyle and driving needs.

In summary, electric vehicles represent a cleaner, more sustainable, and technologically advanced mode of transportation compared to traditional internal combustion engine vehicles. By embracing electric mobility, you're not just upgrading your ride – you're contributing to a greener and more environmentally responsible future for generations to come.

Assessing Your Needs and Budget

As you embark on the journey of purchasing an electric vehicle (EV) in 2024, it's essential to assess your individual needs and financial considerations. By taking the time to evaluate your driving habits, daily commuting distance, and access to charging infrastructure, as well as understanding your budget constraints, you can make a well-informed decision that aligns with your lifestyle and preferences.

Start by reflecting on your typical driving patterns and routines. Consider the distance you travel on a daily basis, including your commute to work or school, errands, and weekend outings. Assessing your driving habits will help you determine the range requirements of your EV, ensuring that it can comfortably meet your daily transportation needs without requiring frequent recharging.

Next, consider the availability and accessibility of charging infrastructure in your area. Determine whether you have access to convenient charging options at home, work, or public charging stations. Understanding your charging options will help you plan for recharging your EV and alleviate any concerns about range anxiety.

In addition to assessing your driving needs, it's crucial to evaluate your budgetary constraints. Take into account the upfront costs of purchasing an EV, including the purchase price, taxes, and registration fees. Consider potential savings on fuel costs, as EVs typically have lower operating costs compared to traditional gasoline-powered vehicles. Factor in any available incentives, such as tax credits, rebates, or discounts offered by government agencies or EV manufacturers.

Explore financing options available for EV purchases, such as loans, leases, or special financing programs offered by automakers or financial institutions. Determine a budget that aligns with your financial goals and obligations, ensuring that you can afford the monthly payments and associated costs of owning an EV.

By carefully assessing your needs and budget, you can narrow down your options and choose the right EV that meets your requirements while staying within your financial means. Whether you prioritize range, charging convenience, or affordability, understanding your needs and budget will empower you to make a confident and informed decision as you transition to electric mobility.

Choosing the Right EV

As you navigate through the plethora of electric vehicle (EV) options available in the market, it's crucial to find the perfect match for your unique lifestyle and preferences. With a diverse range of EV models offering varying features and capabilities, it can be overwhelming to make a decision. Here's how you can streamline the process and choose the right EV for you:

1. **Consider Your Driving Needs**: Start by evaluating your typical driving habits and requirements. If you have a long daily commute or frequently embark on road trips, prioritize EV models with longer driving ranges to ensure you can reach your destinations without worrying about recharging frequently. Conversely, if you primarily use your vehicle for short trips around town, a smaller EV with a more modest range may suffice.

Example: If you commute 50 miles each way to work every day, consider EV models with a range of at least 200 miles to ensure you have ample range for your daily commute without needing to recharge frequently.

2. **Evaluate Charging Infrastructure**: Assess the availability and accessibility of charging infrastructure in your area. If you have access to convenient home charging or a reliable network of public charging stations, you may have more flexibility in choosing an EV with a shorter range. However, if charging options are limited in your area, prioritize EV models with longer ranges or fast-charging capabilities to minimize range anxiety.

Example: If you live in an apartment building without access to home charging and rely on public charging stations, consider EV models with fast-charging capabilities that can replenish the battery quickly when you're on the go.

3. **Compare Features and Amenities**: Explore the features and amenities offered by different EV models to find the right balance of comfort, convenience, and technology. Consider factors such as interior space, infotainment systems, driver assistance features, and available

upgrades. Test drive multiple EVs to experience firsthand how they handle, ride, and feel behind the wheel.

Example: If you prioritize technology and connectivity features, consider EV models with advanced infotainment systems, smartphone integration, and driver assistance features such as adaptive cruise control and lane-keeping assist.

4. **Research Reliability and Ownership Costs**: Investigate the reliability ratings and ownership costs of potential EV models to ensure you're making a sound investment. Look for reviews from current EV owners, reliability ratings from reputable sources, and estimates of long-term ownership costs, including maintenance, repairs, and insurance.

Example: Before making a decision, research reliability ratings and ownership costs for EV models you're considering. Choose a model with a reputation for reliability and reasonable ownership costs to minimize unexpected expenses down the road.

By carefully considering factors such as driving needs, charging infrastructure, features, reliability, and ownership costs, you can narrow down your options and choose the perfect EV that aligns with your lifestyle, preferences, and budget. Whether you prioritize range, charging speed, technology, or affordability, finding the right EV will enhance your driving experience and pave the way for a greener and more sustainable future.

Understanding Incentives

As you embark on your journey to purchase an electric vehicle (EV) in 2024, it's important to be aware of the various government incentives and grants available to EV buyers. These incentives can play a significant role in reducing the upfront cost of purchasing an EV and making electric mobility more accessible to a wider range of consumers. Here's a breakdown of the incentives available in the US, UK, and EU:

United States (US):

Federal Tax Credit: The US federal government offers a tax credit of up to $7,500 for the purchase of qualifying EVs. The amount of the credit varies depending on the vehicle's battery size and range.

Example: If you purchase an EV with a battery capacity of at least 16 kWh, you may be eligible for the full $7,500 tax credit.

State and Local Incentives: Many states and local governments offer additional incentives for EV buyers, such as rebates, tax credits, or reduced registration fees. These incentives vary by location and may be subject to income eligibility requirements.

Example: In California, EV buyers may be eligible for a rebate of up to $2,500 through the Clean Vehicle Rebate Project (CVRP).

United Kingdom (UK):

Plug-In Car Grant: The UK government offers a Plug-In Car Grant of up to £2,500 for eligible EVs with a zero-emission range of at least 70 miles.

Example: If you purchase an EV with a zero-emission range of at least 70 miles, you may be eligible for a grant of up to £2,500.

Vehicle Excise Duty (VED) Exemption: EV owners are exempt from paying Vehicle Excise Duty (road tax) in the UK, saving them hundreds of pounds annually.

Example: As an EV owner in the UK, you won't have to pay any road tax, saving you money each year.

European Union (EU):

EU Clean Vehicles Directive: The EU encourages member states to implement incentives and support measures for the deployment of clean vehicles, including EVs. Incentives may include tax incentives, grants, subsidies, or preferential access to parking and low-emission zones.

Example: In Germany, EV buyers may be eligible for a purchase subsidy of up to €6,000 through the Environmental Bonus program.

Before making your purchase, be sure to research and take advantage of the incentives available in your region. These incentives can help offset the cost of purchasing an EV and make electric mobility more affordable and accessible for you.

Accessing Charging Infrastructure

As you delve into the world of electric vehicle (EV) ownership in 2024, understanding and accessing charging infrastructure is paramount to your overall experience. Here's how you can navigate the landscape of charging options to ensure seamless and convenient charging for your EV:

1. **Public Charging Stations**: Take the time to familiarize yourself with the locations of public charging stations in your area and along your regular routes. Use online maps and mobile apps to locate charging stations and check for availability and compatibility with your EV. Consider factors such as charging speed, connector types, and network coverage when planning your charging stops.

2. **Home Charging Equipment**: Consider installing a home charging station or wall-mounted charger to conveniently recharge your EV overnight or during periods of low demand. Consult with an electrician to assess your home's electrical infrastructure and determine

the most suitable charging solution for your needs. Explore options such as Level 1 chargers (120-volt outlet) for slower overnight charging or Level 2 chargers (240-volt circuit) for faster charging times.

3. **Long-Distance Travel Planning**: If you anticipate embarking on long-distance trips or road trips with your EV, plan your routes carefully to ensure access to fast-charging stations along the way. Research charging networks and service providers that offer reliable and convenient access to fast-charging stations, allowing you to recharge your EV quickly and efficiently during extended journeys.

4. **Charging Networks and Providers**: Explore the various charging networks and service providers available in your region, such as ChargePoint, EVgo, Electrify America, and Tesla Superchargers. Consider joining membership programs or subscription services offered by charging networks to access discounted charging rates, exclusive perks, and enhanced

charging experiences.

5. **Charging Speeds and Connector Types**: Familiarize yourself with the different charging speeds and connector types used in EV charging infrastructure. Understand the differences between Level 1, Level 2, and Level 3 (DC fast charging) chargers, as well as connector standards such as CHAdeMO, CCS (Combined Charging System), and Tesla's proprietary connector. Choose charging stations that are compatible with your EV's charging capabilities to optimize charging speed and efficiency.

6. **Convenient Access to Fast-Charging Stations**: Prioritize locations with convenient access to fast-charging stations, such as shopping centers, rest areas, and highway service plazas. Plan your charging stops strategically to minimize downtime and maximize your driving range, ensuring peace of mind on the road.

By familiarizing yourself with public charging stations, installing home charging equipment, planning for long-distance travel, exploring charging networks, understanding charging speeds and connector types, and prioritizing convenient access to fast-charging stations, you can navigate the charging infrastructure landscape with confidence and enjoy a seamless and enjoyable EV ownership experience.

Considerations for EV Ownership

As you embark on your journey as an electric vehicle (EV) owner in 2024, it's important to address common considerations and questions that may arise regarding EV ownership. By understanding key factors such as range anxiety, battery life, maintenance requirements, and environmental benefits, you can optimize your EV ownership experience and maximize the return on your investment. Here are some considerations to keep in mind:

1. **Range Anxiety**: Range anxiety, or the fear of running out of battery charge before reaching your destination, is a common concern among EV owners. While EVs have made significant advancements in battery technology and driving range, it's important to plan your trips carefully and utilize available charging

infrastructure to alleviate concerns about range anxiety. Adopt energy-efficient driving habits, such as avoiding rapid acceleration and braking, and use tools such as onboard range estimators and navigation systems to plan your routes and charging stops effectively.

2. **Battery Life**: EV batteries are designed to last for many years, but it's natural to have questions about battery longevity and performance. Modern EV batteries are typically covered by warranties that guarantee a certain level of performance and capacity retention over a specified period, providing peace of mind for EV owners. To prolong the life of your EV battery, follow recommended charging practices, such as avoiding frequent deep discharges and extreme temperature conditions, and adhere to manufacturer guidelines for battery maintenance and care.

3. **Maintenance Requirements**: One of the advantages of EV ownership is reduced maintenance compared to traditional internal combustion engine vehicles. EVs have fewer

moving parts and require less frequent maintenance, resulting in lower operating costs over time. While routine maintenance tasks such as tire rotations, brake inspections, and cabin air filter replacements are still necessary, EVs eliminate the need for oil changes, spark plug replacements, and other traditional maintenance tasks associated with ICE vehicles.

4. **Environmental Benefits**: By driving an EV, you're not only reducing your carbon footprint but also contributing to cleaner air and a healthier environment. EVs produce zero tailpipe emissions, helping to improve air quality and mitigate the impacts of climate change. Additionally, EVs are more energy-efficient than ICE vehicles, as they convert a higher percentage of stored energy into forward motion, reducing overall energy consumption and dependence on fossil fuels.

5. **Energy-Efficient Driving Habits**: Adopt energy-efficient driving habits to maximize your EV's range and efficiency. Avoid aggressive acceleration and braking, maintain a steady

speed, and anticipate traffic patterns to minimize energy consumption. Use regenerative braking to capture and recapture energy during deceleration, further extending your EV's range and reducing wear on the braking system.

6. **Follow Maintenance Practices**: Follow manufacturer recommendations for routine maintenance and care to ensure optimal performance and longevity of your EV. Schedule regular service appointments for inspections, fluid checks, and software updates, and address any issues or concerns promptly to prevent potential problems from escalating.

By addressing these common considerations and questions related to EV ownership, you can confidently embrace electric mobility and enjoy the numerous benefits of driving an EV. With proper planning, maintenance, and environmental stewardship, you can optimize your EV ownership experience and contribute to a more sustainable future for generations to come.

Chapter 14: Future Trends and Technology

As we peer into the future of electric vehicles (EVs), it's evident that the trajectory of this transformative technology is poised to reshape our transportation systems in profound ways. Here's a glimpse into some of the future trends and technologies that are expected to drive the evolution of the EV market:

Advancements in Battery Technology

Solid-State Batteries: Among the most promising advancements are solid-state batteries, heralded for their potential to redefine the capabilities of EVs. These batteries replace the liquid electrolyte found in traditional lithium-ion batteries with a solid electrolyte, offering several advantages. Firstly, solid-state batteries promise higher energy density, meaning they can store more energy within the same volume, potentially extending the driving range of EVs. Secondly, they boast faster charging times, addressing one of the main concerns for EV owners. With solid-state batteries, EVs could be recharged in a matter of minutes, comparable to refueling a conventional vehicle. Thirdly, solid-state batteries offer improved safety due to their stable chemistry, reducing the risk of thermal runaway and enhancing overall reliability.

Alternative Battery Chemistries: Research into alternative battery chemistries, such as lithium-sulfur and lithium-air, represents another frontier in battery innovation. These chemistries hold the promise of even greater energy storage capacity and efficiency compared to conventional lithium-ion batteries. Lithium-sulfur batteries, for example, have the potential to surpass the energy density of lithium-ion batteries while using more abundant and cheaper materials. Similarly, lithium-air batteries could offer significantly higher energy densities, paving the way for EVs with extended driving ranges and reduced charging frequency. While challenges remain in terms of stability, lifespan, and scalability, ongoing research efforts are driving progress towards commercialization.

Impact on the EV Industry: The widespread adoption of next-generation battery technologies has the potential to revolutionize the EV industry. With solid-state batteries and alternative chemistries, EV manufacturers can overcome the limitations of current lithium-ion batteries, offering consumers vehicles with greater range, faster charging times, and enhanced safety. These advancements could accelerate the mass adoption of EVs by addressing key concerns such as range anxiety and charging infrastructure limitations. Moreover, the improved energy density and efficiency of next-generation batteries could open doors to new applications beyond passenger vehicles, including electric trucks, buses, and even aircraft, transforming entire sectors of the transportation industry.

Challenges and Opportunities: While the future of battery technology holds immense promise, it is not without its challenges. Scaling up production, ensuring cost competitiveness, and addressing concerns related to raw material availability and environmental impact are among the hurdles that must be overcome. Additionally, advancements in battery technology must be accompanied by developments in charging infrastructure and grid integration to support the widespread adoption of EVs. Despite these challenges, the potential benefits of next-generation battery technologies are vast, offering a cleaner, more efficient, and sustainable future for transportation.

Autonomous Driving and Connectivity

The fusion of electric vehicles (EVs) with autonomous driving technology and connectivity represents a paradigm shift in the future of mobility. Here's a detailed exploration into the potential impact of this convergence:

Enhanced Safety and Efficiency: One of the primary promises of autonomous driving technology is the potential to enhance safety on the roads. By eliminating human error, which is a leading cause of accidents, autonomous EVs can significantly reduce the risk of collisions and fatalities. Advanced sensors, cameras, and artificial intelligence algorithms enable these vehicles to perceive and respond to their surroundings with unparalleled precision and speed. Moreover, autonomous EVs can optimize driving patterns and routes, leading to improved traffic flow, reduced congestion, and lower energy consumption.

Convenience and Mobility-as-a-Service: The integration of autonomous EVs into mobility-as-a-service (MaaS) platforms holds the potential to revolutionize urban transportation. Imagine summoning a self-driving EV with a tap on your smartphone, which arrives at your doorstep precisely when you need it. These autonomous EVs can be seamlessly integrated into public transit networks, offering first-mile and last-mile connectivity and bridging gaps in existing transportation systems. By providing convenient, on-demand mobility solutions, autonomous EVs can reduce the need for private car ownership, leading to more sustainable and efficient urban mobility ecosystems.

Transforming the Passenger Experience: Autonomous EVs will redefine the passenger experience, transforming vehicles into mobile living spaces, workspaces, and entertainment hubs. With hands-free driving capabilities, passengers can reclaim valuable time that would otherwise be spent behind the wheel, engaging in productive activities or leisure pursuits during their journeys. Interior designs can be optimized for comfort and convenience, with modular seating arrangements, advanced infotainment systems, and connectivity features that keep passengers connected and entertained on the go.

Regulatory and Technological Challenges: Despite the transformative potential of autonomous EVs, several challenges must be addressed before widespread adoption can occur. Regulatory frameworks governing autonomous driving vary by jurisdiction and must evolve to address safety, liability, and ethical considerations. Additionally, technological hurdles, such as ensuring robust cybersecurity and addressing edge cases and unpredictable scenarios, require ongoing research and development efforts. Moreover, public acceptance and trust in autonomous technology are crucial for its successful adoption, necessitating transparent communication, education, and demonstration of the safety and reliability of autonomous EVs.

Future Implications: The convergence of EVs with autonomous driving technology and connectivity has far-reaching implications for society, the economy, and the environment. From reducing traffic accidents and emissions to revolutionizing urban mobility and enhancing the passenger experience, autonomous EVs offer a glimpse into a future of cleaner, safer, and more efficient transportation. By embracing innovation, collaboration, and responsible deployment, we can unlock the full potential of autonomous EVs and shape a sustainable and equitable mobility landscape for generations to come.

Vehicle-to-Grid Integration

The integration of vehicle-to-grid (V2G) technology represents a groundbreaking advancement in the symbiotic relationship between electric vehicles (EVs) and the electric grid. Here's a closer look at the potential benefits and implications of V2G integration:

Optimizing Energy Usage: V2G technology enables EV batteries to function not only as energy consumers but also as energy providers, allowing for bidirectional flow of electricity between the vehicle and the grid. During times of low demand, EV batteries can store excess renewable energy generated by sources like solar and wind, effectively acting as mobile energy storage units. Conversely, during periods of high demand or when renewable energy generation is limited, EV batteries can discharge stored energy back into the grid, helping to alleviate strain on the electricity network.

Reducing Peak Demand: One of the key advantages of V2G integration is its ability to reduce peak demand on the grid. By allowing EVs to participate in grid balancing services, such as frequency regulation and peak shaving, V2G systems can help smooth out fluctuations in electricity demand, thereby enhancing grid stability and reliability. This can lead to more efficient use of existing infrastructure and potentially reduce the need for costly investments in grid expansion and upgrades.

Facilitating Renewable Energy Integration: V2G technology plays a crucial role in facilitating the integration of renewable energy sources into the grid. By enabling EV batteries to store surplus renewable energy when generation exceeds demand, V2G systems help address the intermittency and variability of renewable energy resources. This, in turn, enhances the reliability and resilience of the grid while reducing dependence on fossil fuels and mitigating greenhouse gas emissions.

Enhancing Grid Resilience: The collective power of EV batteries harnessed through V2G systems can significantly enhance grid resilience, particularly in the face of natural disasters, grid outages, or other emergencies. EVs equipped with V2G capabilities can serve as backup power sources for critical infrastructure, emergency services, and residential homes, providing essential electricity during times of need. This decentralized approach to energy storage and distribution can improve overall grid reliability and resilience to unforeseen events.

Unlocking New Revenue Streams: V2G integration creates new opportunities for EV owners to monetize their vehicles' battery capacity. By participating in grid services and providing ancillary services to utilities and grid operators, EV owners can earn incentives, payments, or credits for the energy they supply to the grid. This not only helps offset the cost of EV ownership but also incentivizes the adoption of clean transportation and renewable energy technologies.

Challenges and Considerations: Despite its potential benefits, V2G integration presents several challenges and considerations. These include technological interoperability, regulatory frameworks, cybersecurity, and privacy concerns. Additionally, V2G implementation requires coordination among stakeholders, including automakers, utilities, regulators, and consumers, to ensure seamless integration and maximize the benefits for all parties involved.

Future Outlook: The future of V2G integration holds immense promise for transforming the energy landscape, promoting energy sustainability, and advancing the transition to a cleaner, more resilient grid. With continued innovation, collaboration, and investment, V2G technology has the potential to revolutionize the way we generate, distribute, and consume electricity, paving the way for a more sustainable and resilient energy future.

Wireless Charging Infrastructure

The advent of wireless charging infrastructure represents a significant leap forward in the evolution of electric vehicles (EVs) and the charging experience.

Let's delve deeper into the potential implications and benefits of wireless charging technology:

Convenience and Seamless Charging Experience: Wireless charging technology eliminates the need for physical cables or plugs, offering EV owners a convenient and hassle-free charging experience. With wireless charging pads installed in parking spaces, garages, and other designated areas, EVs can recharge their batteries simply by parking over the charging pad, without the need to manually plug in. This seamless process makes EV ownership more convenient and accessible to a wider range of users, including those with limited mobility or physical disabilities.

Widespread Adoption and Efficiency: As wireless charging infrastructure becomes more widespread and efficient, EV owners can expect to recharge their vehicles effortlessly at various locations, including home, workplaces, shopping centers, and public parking lots. The integration of wireless charging technology into urban infrastructure, such as streetlights and sidewalks, could further enhance accessibility and convenience for EV users.

Additionally, advancements in wireless charging efficiency and power transfer capabilities are making this technology increasingly viable for commercial and residential applications.

Dynamic Wireless Charging: One of the most promising applications of wireless charging technology is dynamic wireless charging systems embedded in roadways. These systems utilize inductive charging technology to wirelessly transfer power to EVs as they drive over specially equipped road surfaces. Dynamic wireless charging has the potential to extend the driving range of EVs by continuously replenishing their batteries while in motion, effectively eliminating the need for dedicated charging stations and range anxiety. This innovation could revolutionize long-distance travel and enable the widespread adoption of electric mobility on a global scale.

Environmental and Aesthetic Benefits: Wireless charging infrastructure offers environmental benefits by reducing the reliance on traditional fossil fuels and lowering greenhouse gas emissions associated with transportation. By enabling EVs to recharge using renewable energy sources, such as solar or wind power, wireless charging technology contributes to a cleaner and more sustainable transportation ecosystem. Furthermore, the integration of wireless charging pads into urban environments can enhance the aesthetic appeal of streetscapes and public spaces, promoting the adoption of EVs as a stylish and modern transportation solution.

Challenges and Considerations: Despite its potential benefits, wireless charging technology faces several challenges and considerations, including standardization, interoperability, efficiency, and cost. Ensuring compatibility and uniformity across different wireless charging systems is essential to enable widespread adoption and seamless integration into existing infrastructure. Moreover, optimizing wireless charging efficiency and minimizing energy losses are critical for maximizing the environmental and economic benefits of this technology. Addressing these challenges will require collaboration among stakeholders, including automakers, technology developers, policymakers, and utilities, to drive innovation and overcome barriers to adoption.

Future Outlook: The future of wireless charging infrastructure is promising, with ongoing advancements in technology, infrastructure deployment, and regulatory support driving its continued evolution and adoption. As wireless charging becomes more efficient, accessible, and cost-effective, it has the potential to revolutionize the way we power and recharge electric vehicles, paving the way for a cleaner, greener, and more sustainable transportation future.

Environmental Sustainability

The future of electric vehicles (EVs) is closely intertwined with environmental sustainability, with concerted efforts underway to mitigate the carbon footprint associated with all stages of an EV's lifecycle. Here's a closer look at the initiatives and innovations driving environmental sustainability in the realm of EVs:

Responsibly Sourcing Raw Materials: Manufacturers are increasingly focusing on responsibly sourcing raw materials essential for EV production, such as lithium, cobalt, and nickel. This involves implementing ethical mining practices, reducing environmental impacts, and addressing labor concerns in material supply chains. By prioritizing sustainable sourcing practices, manufacturers aim to minimize the environmental and social impacts associated with raw material extraction.

Energy-Efficient Production Processes: Efforts are underway to minimize energy consumption and reduce greenhouse gas emissions during the production of EVs. Manufacturers are investing in energy-efficient manufacturing processes, renewable energy sources, and innovative technologies to lower the carbon intensity of EV production. By embracing cleaner energy alternatives and optimizing production efficiency, manufacturers seek to mitigate the environmental footprint of EV manufacturing.

Lifecycle Environmental Impact Reduction: The adoption of circular economy principles is central to reducing the lifecycle environmental impact of EVs. This includes initiatives such as battery recycling and reuse, extending the lifespan of EV components, and minimizing waste generation throughout the vehicle's lifecycle. By implementing sustainable end-of-life strategies, manufacturers aim to recover valuable materials from retired EVs, reduce reliance on virgin resources, and minimize the environmental burden of disposal.

Battery Recycling and Reuse: Battery recycling and reuse are critical components of sustainable EV management. As EV battery technology advances, efforts are underway to develop efficient recycling processes that recover valuable materials such as lithium, cobalt, and nickel for reuse in new batteries. By closing the loop on battery materials and minimizing resource extraction, battery recycling contributes to the circularity and sustainability of the EV ecosystem.

Regulatory Support and Industry Collaboration: Governments, regulatory bodies, and industry stakeholders are collaborating to establish policies and standards that promote environmental sustainability in the EV sector. This includes regulations mandating eco-friendly manufacturing practices, incentives for battery recycling and reuse, and targets for reducing greenhouse gas emissions throughout the EV lifecycle. By aligning regulatory frameworks with sustainability objectives, policymakers aim to accelerate the transition to a cleaner and more sustainable transportation system.

Consumer Awareness and Education: Educating consumers about the environmental benefits of EVs and sustainable driving practices is essential for fostering widespread adoption and support. Initiatives to raise awareness about the environmental impact of transportation choices, promote energy-efficient driving habits, and highlight the sustainability features of EVs can empower consumers to make informed decisions that align with their environmental values.

Future Outlook: The future of EVs hinges on continued innovation, collaboration, and commitment to environmental sustainability. By leveraging advances in technology, embracing circular economy principles, and prioritizing environmental stewardship, the EV industry can contribute to a greener, more sustainable future for transportation. With ongoing efforts to reduce carbon emissions, conserve resources, and minimize environmental impact, EVs are poised to play a central role in transitioning towards a more sustainable mobility paradigm.

Policy and Regulatory Support

Policy and regulatory support are pivotal in fostering the widespread adoption of electric vehicles (EVs) and advancing the transition to sustainable transportation systems. Here's an in-depth exploration of the importance of continued policy and regulatory backing:

Consumer Incentives and Support: Governments worldwide are implementing various incentives and support mechanisms to encourage consumers to embrace EVs. These incentives may include tax credits, rebates, grants, and subsidies aimed at reducing the upfront costs of purchasing EVs and making them more financially appealing to consumers. Additionally, investment in charging infrastructure and public transportation systems enhances accessibility and convenience for EV users, further incentivizing adoption.

Emissions Regulations and Vehicle Standards: Stringent emissions regulations and vehicle standards play a crucial role in driving automakers to innovate and invest in cleaner technologies. By setting targets for reducing greenhouse gas emissions and imposing emission limits on vehicles, governments create a regulatory environment that incentivizes the development and adoption of EVs and other low-emission vehicles. Vehicle standards also promote safety, quality, and performance standards for EVs, ensuring that they meet rigorous requirements before entering the market.

Infrastructure Investments: Governments are investing in the expansion and enhancement of EV charging infrastructure to support the growing fleet of EVs on the roads. This includes funding for the installation of charging stations in public spaces, workplaces, residential areas, and along highways, as well as incentives for private entities to invest in charging infrastructure development. Robust charging infrastructure is critical for addressing range anxiety, facilitating long-distance travel, and increasing consumer confidence in EVs.

Research and Development Funding: Public investment in research and development (R&D) initiatives is essential for driving innovation in EV technology and advancing the state of the art. Governments provide funding for R&D projects focused on improving battery technology, enhancing charging infrastructure, optimizing vehicle efficiency, and addressing other key challenges in the EV ecosystem. By supporting collaborative research efforts between academia, industry, and government agencies, policymakers foster innovation and drive progress in the EV sector.

International Cooperation and Collaboration: Collaboration among governments, international organizations, and industry stakeholders is essential for harmonizing policies, sharing best practices, and addressing common challenges in the global transition to EVs. International cooperation facilitates the exchange of knowledge, expertise, and resources, enabling countries to learn from each other's experiences and accelerate the adoption of EVs on a global scale. By working together towards common goals, countries can overcome barriers to EV adoption and create a more sustainable future for transportation.

Public Awareness and Education: Public awareness campaigns and education initiatives play a vital role in promoting EV adoption and garnering public support for policy measures aimed at advancing sustainable transportation. Governments invest in outreach programs, informational resources, and educational campaigns to raise awareness about the benefits of EVs, dispel myths and misconceptions, and encourage informed decision-making among consumers. By fostering a culture of sustainability and environmental consciousness, policymakers can drive positive attitudes towards EVs and pave the way for their widespread acceptance.

Future Outlook: The future of EVs hinges on continued policy and regulatory support from governments worldwide. By implementing incentives, regulations, and investments that promote EV adoption, policymakers can accelerate the transition to sustainable transportation and realize the environmental, economic, and social benefits of electric mobility. With concerted efforts and collaborative action, governments can create an enabling environment that fosters innovation, investment, and growth in the EV sector, paving the way for a cleaner, greener, and more sustainable transportation future.

As these future trends and technologies converge, the landscape of electric mobility will continue to evolve, offering new opportunities and challenges for stakeholders across the transportation ecosystem. From advancements in battery technology and autonomous driving to the proliferation of wireless charging infrastructure and environmental sustainability initiatives, the future of EVs holds the promise of a cleaner, smarter, and more sustainable transportation future for generations to come.

Appendix I : Electric Vehicle Manufacturers and Models

1. Tesla, Inc.
- **Model S**: A luxury sedan with long-range capabilities and high-performance features.
- **Model 3**: A compact sedan designed for mass-market appeal, offering a balance of range, performance, and affordability.
- **Model X**: An all-electric SUV with falcon-wing doors and advanced safety features.
- **Model Y**: A compact SUV built on the same platform as the Model 3, offering versatility and practicality.

2. Chevrolet (General Motors)
- **Chevrolet Bolt EV**: A compact hatchback with an affordable price point and a long electric range.

3. Nissan Motor Corporation
- **Nissan LEAF**: One of the best-selling electric vehicles globally, offering practicality, efficiency, and a comfortable driving

experience.

4. Audi AG
- **Audi e-tron**: A premium electric SUV with a luxurious interior, advanced technology features, and impressive performance.

5. BMW AG
- **BMW i3**: A compact electric hatchback with distinctive styling and an emphasis on sustainability.
- **BMW iX3**: An electric version of the popular X3 SUV, offering all-electric driving with BMW's signature performance and luxury.

6. Hyundai Motor Company
- **Hyundai Kona Electric**: A subcompact SUV with a long electric range and a versatile interior.
- **Hyundai Ioniq Electric**: A compact electric hatchback available with a range of efficient features and technologies.

7. Kia Corporation
- **Kia Soul EV**: A quirky and practical electric vehicle with ample cargo space and a comfortable interior.

8. Ford Motor Company
- **Ford Mustang Mach-E**: An all-electric SUV inspired by the iconic Mustang, offering performance and style.

9. Volkswagen Group
- **Volkswagen ID.4**: A compact electric SUV with a sleek design and advanced driver-assistance features.

10. Rivian Automotive, Inc.
- **R1T**: An all-electric pickup truck designed for adventure and off-road capabilities.
- **R1S**: An electric SUV built on the same platform as the R1T, offering versatility and performance.

11. Lucid Motors
- **Lucid Air**: A luxury electric sedan with cutting-edge technology, luxurious amenities, and impressive performance.

12. Polestar (Volvo Cars)
- **Polestar 2**: A premium electric sedan with a focus on design, performance, and

sustainability.

13. Porsche AG
- **Porsche Taycan**: A high-performance electric sports car with Porsche's signature driving dynamics and craftsmanship.

14. Jaguar Land Rover
- **Jaguar I-PACE**: A luxury electric SUV with sleek styling, dynamic performance, and advanced technology features.

15. Mercedes-Benz (Daimler AG)
- **Mercedes-Benz EQC**: A luxury electric SUV offering comfort, performance, and cutting-edge technology.

This list provides an overview of some of the major electric vehicle manufacturers and some of the models they offer. It's important to consult manufacturers' websites or contact local dealerships for the most up-to-date information on pricing, availability, and specifications.

Official Websites:

1. Tesla, Inc.
    - Official Website: www.tesla.com

2. Chevrolet (General Motors)
    - Official Website: www.chevrolet.com/electric

3. Nissan Motor Corporation
    - Official Website: www.nissanusa.com/vehicles/electric-cars

4. Audi AG
    - Official Website: www.audi.com/en/experience-audi/models-and-technology/efficiency-and-emissions/audi-electric-cars.html

5. BMW AG
    - Official Website: www.bmwusa.com/electric.html

6. Hyundai Motor Company

- Official Website: www.hyundaiusa.com/us/en/vehicles

7. Kia Corporation
   - Official Website: www.kia.com/us/en/electric-vehicles/

8. Ford Motor Company
   - Official Website: www.ford.com/electric/

9. Volkswagen Group
   - Official Website: www.vw.com/electric-concepts/

10. Rivian Automotive, Inc.
    - Official Website: www.rivian.com

11. Lucid Motors
    - Official Website: www.lucidmotors.com

12. Polestar (Volvo Cars)
    - Official Website: www.polestar.com

13. Porsche AG
    - Official Website:

www.porsche.com/usa/models/taycan/taycan-models/

14. Jaguar Land Rover
    - Official Website: www.jaguarusa.com/electrification.html

15. Mercedes-Benz (Daimler AG)
    - Official Website: www.mercedes-benz.com/en/innovation/eq/

About the Author

Etienne Psaila, an accomplished author with over two decades of experience, has mastered the art of weaving words across various genres. His journey in the literary world has been marked by a diverse array of publications, demonstrating not only his versatility but also his deep understanding of different thematic landscapes. However, it's in the realm of automotive literature that Etienne truly combines his passions, seamlessly blending his enthusiasm for cars with his innate storytelling abilities.

Specializing in automotive and motorcycle books, Etienne brings to life the world of automobiles through his eloquent prose and an array of stunning, high-quality color photographs. His works are a tribute to the industry, capturing its evolution, technological advancements, and the sheer beauty of vehicles in a manner that is both informative and visually captivating.

A proud alumnus of the University of Malta, Etienne's academic background lays a solid foundation for his meticulous research and factual accuracy. His education has not only enriched his writing but has also fueled his career as a dedicated teacher. In the classroom, just as in his writing, Etienne strives to inspire, inform, and ignite a passion for learning.

As a teacher, Etienne harnesses his experience in writing to engage and educate, bringing the same level of dedication and excellence to his students as he does to his readers. His dual role as an educator and author makes him uniquely positioned to understand and convey complex concepts with clarity and ease, whether in the classroom or through the pages of his books.

Through his literary works, Etienne Psaila continues to leave an indelible mark on the world of automotive literature, captivating car enthusiasts and readers alike with his insightful perspectives and compelling narratives.

He can be contacted personally on etipsaila@gmail.com